W9-BVX-752

RESOURCE MATERIALS FOR

ENVIRONMENTAL MANAGEMENT AND EDUCATION

RESOURCE MATERIALS FOR ENVIRONMENTAL MANAGEMENT AND EDUCATION

William H. Matthews

with

Joseph C. Perkowski

Siu Kee So

Frederick A. Curtis

William F. Martin

The MIT Press
Cambridge, Massachusetts, and London, England

PUBLISHER'S NOTE

This format is intended to reduce the cost of publishing certain works in book form and to shorten the gap between editorial preparation and final publication. The time and expense of detailed editing and composition in print have been avoided by photographing the text of this book directly from the author's typescript.

Copyright © 1976 by
The Massachusetts Institute of Technology

All rights reserved. No part of this book may be reproduced in any form or by any means, electronic or mechanical, including photocopying, recording, or by any information storage and retrieval system, without permission in writing from the publisher.

Printed in the United States of America

Matthews, William Henry, 1942-
Resource materials for environmental management and education.

1. Environmental policy--Study and teaching.
2. Environmental policy--Bibliography. I. Title.
HC79.E5M365 301.31'07 75-28368
ISBN 0-262-13118-8

301.3107
M 442 r

To

Maurice F. Strong

124036

CONTENTS

PREFACE

As concern about environmental problems has increased over the past few years, there has been a corresponding concern about the lack of professional resources available to attack these problems and to prevent their occurrence in the future. Almost every group that has studied this area--whether public, private, or governmental--has reached the same conclusion.

The statement of this conclusion which provided the impetus for this book was in the Report of the Secretary-General on the Action Plan for the 1972 United Nations Conference on the Human Environment where he observed that it was found for "virtually every area of environmental concern" that "there are very few professionals with the training and experience needed to understand and manage multidisciplinary relationships, particularly those involving the common ground shared by scientists and technologists, on the one hand, and the political decision-makers, on the other."

Following the U.N. Conference, the General Assembly established the United Nations Environment Programme (UNEP) which is headquartered in Nairobi, Kenya and is charged with stimulating, initiating, and coordinating environmental activities among nations and within the U.N. system. One of the first project contracts supported by UNEP was with M.I.T. to begin developing some of the resource materials on which future educational programs could be based. The initial outline for the project that produced this book was developed by Mr. Maurice F. Strong, Executive Director of UNEP, and Professors Carroll L. Wilson of M.I.T. and me.

Collaborative arrangements were made between staff at M.I.T. and the Centre d'Études Industrielles (C.E.I.)--the Center for Education in International Management--in Geneva, Switzerland. Grants were made by the Edna McConnell Clark Foundation to M.I.T. and to C.E.I. to conduct complementary studies designed to produce resource materials and a "market survey" of the environmental management functions in multinational corporations. This book compiles and presents much of the work done at M.I.T. on this project.

In the course of this project, we have worked with C.E.I. to identify the educational needs of "environmental managers" in multinational corporations and have begun work with UNEP to determine how these needs can be identified for policy planners in developing countries. The explication of these needs is the next step of this process. When that is done, the materials in this book can be expanded and can be used more effectively to provide a framework and substance for responding to these needs.

The processes of need identification and experimental workshops and seminars that have been initiated as part of this project comprise a substantial part of the work done and the ultimate results that will be achieved. Thus this book presents only part of our work--that part which can be compiled and summarized in written form at this time. The processes that we have helped set in motion should continue to produce results that will ultimately lead to better environmental management and to focused curriculum development for professional education in environmental management at many levels of public and private sector decision-making.

The work at M.I.T. has been conducted by a staff of four graduate students and myself. We relied very heavily on interviews and interactions with over one hundred professionals and scientists from a dozen universities and several public and private agencies and we are very grateful to them for their ideas, suggestions, and insights. Without their guidance, Part II could never have been compiled.

Each of the staff worked for different periods of time and had different responsibilities on the project. While there was much interaction and exchange of ideas it is possible to identify the major contributors of certain sections of this book. Mr. Joseph C. Perkowski worked on the project the longest and was primarily responsible for the sections on "Actor/Role Interactions" in Parts I and II and "Modeling" and "Monitoring" in Part II. He also contributed to the overall integration of the various sections and in writing the introduction to each of the "bodies of knowledge." Mr. Siu Kee So had primary responsibility for the sections on "Environmental Assessment Methodology" in Parts I and II and on "Values and Perceptions", "Environmental Effects", and "Environmental Indicators" in Part II. Mr. Frederick A. Curtis developed the sections on "Ecology", "Environmental Law", and "Administrative Processes" in Part II, and made some contribution to the section on "Environmental Effects". Mr. William F. Martin contributed to the project during its first stages and developed the sections on "Growth" and "Economics" in Part II.

I would like to give special acknowledgement and thanks to Mrs. Annette Pearson who typed the manuscript for this book and worked with The M.I.T. Press to produce a camera-ready copy.

While none of us would claim that this work is comprehensive or definitive, we do hope that it will be of some assistance to those who want to learn more about environmental management or who want to develop educational programs in this area. There is much more work to be done and many more references that could be cited, especially those in other countries. We feel that we can aid this development process by making this contribution in a timely manner rather than continuously refining and supplementing it before publication.

William H. Matthews
June 1975

PART I

CONCEPTUAL APPROACH TO PROFESSIONAL EDUCATION IN ENVIRONMENTAL MANAGEMENT

INTRODUCTION

A few years ago when the "environment" began to appear on the lists of national and international priorities, there seemed to be no experts to whom one could turn to get answers or even adequate definitions of the problems. Now almost the opposite situation exists. Literally everyone--from layperson to scientist, from engineer to philosopher, from lawyer to anthropologist--asserts that he has something vital to contribute to defining and resolving environmental concerns, and they are all correct.

Through this evolution it has become increasingly clear that the most critical environmental problems of our societies do not correspond neatly to the areas of differentiation and specialization that we have developed to advance knowledge and skills in the dozens of engineering, natural science, and social science disciplines, and the various professions. It is also apparent that the past and present functions and roles in the public and private sectors for directing and managing the various activities of man are often not very helpful in supporting comprehensive and rational approaches to environmental management.

Thus, a long term strategy for more effective management of the earth's resources and environmental systems must include the following three elements: a reorganization of our extensive substantive knowledge in these areas, the development of new functions and roles in the public and private sectors, and the creation of new educational programs and institutions to meet the critical needs of a wide variety of persons who analyze environmental problems or who make decisions concerning the environment.

There are essentially five major steps required for implementing this strategy: 1) development of a conceptual framework for considering environmental management processes; 2) development of outlines of the substantive bodies of knowledge that are relevant to environmental management; 3) description of the present and needed functions and roles that comprise environmental management; 4) relating the above to each other and deriving criteria for developing new educational programs; and 5) development of those new programs.

This book presents our results thus far on each of these steps. The following sections of Part I present the major elements and concepts involved in step 1. Part II presents the outlines for step 2. The section in Part I on "The Study of Actor/Role Interactions" presents our very preliminary efforts at steps 3 and 4.

Step 5, the development of new programs, has been begun by developing subjects at M.I.T. and special seminars at the Centre d'Études Industrielles (C.E.I.)--Center for Education in Industrial Management--in Geneva. The C.E.I. programs focus on the educational needs of "environmental managers" in multinational corporations and in the public sector of several countries. The materials in this book were used in preparing the first major seminar in the C.E.I. series in January 1975. Forty-two senior professionals from nineteen countries and representing twenty companies and fifteen governmental and international organizations participated. The educational experience was very well received and judged to be so productive that many of the participants formed an International Association of Environmental Coordinators to foster and expand exchange of experience and information.

The experiences with these materials and concepts at C.E.I. and M.I.T. suggest that they can be very useful to professionals and scientists involved in environmental management processes. The outlines and bibliographies are not definitive, however, and much more work needs to be done in developing the subject areas and in expanding the references particularly to include articles and books published outside of the United States. Nor are these materials operational for instituting educational programs--they must be combined with an understanding of the needs and learning styles of educational "clients", an innovative approach to creating meaningful educational experiences, and a commitment to integrate all of these with the substantive knowledge into a series of educational programs.

ENVIRONMENTAL MANAGEMENT PROCESSES

There are really no formalized comprehensive environmental management processes in the sense that there are formal and hierarchical structures for managing major enterprises in the public and private sectors. There are instead a myriad of individual and collective decisions by persons, groups, and organizations throughout society that result in impacts--both positive and negative--on environmental resources. Taken together these decisions and the interactions among those involved constitute a process--a process that in effect results in the management (or mismanagement) of the environmental resources of the society.

The study of environmental management involves obtaining an understanding of these social and political processes and of the natural environmental elements and processes that are affected by societal decision-making in all of its varied forms. Such study encompasses the understanding of all of nature (the environment) and of all of man's institutions and means of collective decision-making (management). Proficiency is not possible in this "field" because there is not enough collective knowledge, theory, or wisdom to define or to describe what would have to be mastered.

It should, however, be possible for those with real concern and responsibility for environmental resources to obtain some better understanding than they have now of what would be involved in making these management processes more rational and less random and more responsive to environmental criteria rather than principally to economic or expediency or other criteria. This project and the related work at C.E.I. are steps toward finding ways to provide a somewhat better perspective on the processes to aid those with this goal.

The next section of this Part and all of Part II are devoted to a description of some of the bodies of knowledge that will help managers identify the substantive information they will need for this task. In this section a conceptual framework of decision-making steps will be presented that should help in the identification of the major elements in the diverse processes that make up environmental management.

A useful starting point in this understanding is to consider why we have environmental problems. It is essential to realize that the disruption of the environment is not a goal or an end of any specific activity. It is rather the by-product of activities or the means through which activities are conducted. These activities have in general been initiated to fulfil very important human needs such as food, shelter, security, and "amenities." In the process of filling these needs (and wants), man in his inventiveness has determined how to use and manipulate his environment. Sometimes he has used environmental resources directly (such as wood or oil) and sometimes he has used them indirectly (such as for a free disposal system).

Recently societies have begun recognizing that part of the costs we are paying for those goods and services we "need" involve the loss of some environmental resources. This has created the conflict. It is a conflict between the desirability of continuing to obtain what we "need" or the way we obtain it versus the desirability of not despoiling, degrading, or destroying finite environmental resources. It is not a choice of giving up something "bad" such as polluting--it is a choice of giving up or of paying more for a good or service that we had previously decided we wanted and that someone has undertaken to provide for us.

Thus in the overall context there are no pure "villains". There are, however, persons who meet some of our needs while at the same time depriving us (or others) of other things which we or they also value. Resolution of these conflicts that involve change therefore can deeply affect the legitimate interests of some segment of the society who now find part of the consequences of their socially desirable activities on a new list of undesirable costs.

This is the great dilemma of resolving conflicts in environmental management. If we decide that pollution of a certain river is undesirable, we cannot simply close the polluting plant for its business is not to pollute but to create products that people want to buy and jobs for people who want to work. If we decide we are using too many trees, we cannot simply tell lumber companies to quit cutting trees because their business is not to kill trees but to provide wood and paper products for an enormous number of activities throughout the entire society.

Thus every decision in environmental management must be made in the broader context of meeting all societal needs. We must not only decide what kind of environment we want, but we must also decide what we are willing to give up or to do in addition so that we can effect the change in the status quo that will generally be required to produce a change in the environment. Such decision-making is incredibly difficult under the best of circumstances. It becomes almost intractable in complex, pluralistic, heterogeneous societies when scientific information is incomplete and often speculative.

Environmental management as an operational concept might best be defined as describing the actions of those involved in decision-making who make a deliberate and systematic attempt to understand how their activities affect the environment and how their concerns about the environment will affect their activities. These persons would strive to make explicit the values involved, the objective information available, and the trade-offs that would be necessary. They would then attempt to resolve the issues within the relevant public and private decision-making spheres. Such actions are in contrast either to those who pursue single-mindedly one set of societal objectives with no consideration of environmental values or those who pursue narrowly defined "conservation" goals with no attempt to analyze the major changes that would occur in society as a consequence of radical changes in the status quo of human activities and societal development.

Thus, environmental management is a positive concept--not a negative "halt progress" or "back to nature" concept. Some of the key aspects involved are:

--Identification of the "needs" and "wants" of man in his individual and social development

--Identification of the resources that are needed for such development

--Identification of how meeting some needs and wants will inhibit or increase the ability to meet others as a result of the destruction, degradation, or enhancement of resources

--Identification of the actors who have a stake in the conflicts that may arise between meeting needs and protecting resources

--Resolution of these conflicts

There are two major elements involved in analyzing environmental management processes: 1) the role of values or purely subjective judgements in decision-making, and 2) the role of scientific information or purely objective judgements. Because environmental management involves both people and the natural and physical environment, these two elements are present in every decision. The people supply the values and the "laws of nature" create the need for a scientific objective understanding of cause-effect relation in the physical world.

Figure 1 presents a very simplified idealized diagram of the decision-making steps in environmental management processes. The real processes are much more complicated and certainly do not follow such an orderly and logical set of steps. This framework is however useful for isolating major decision steps and determining the educational needs of decision-makers for that step.

The inputs, outputs, and processes of each step will, of course, vary considerably among problem areas and depending on the types of actors who have the decision-making responsibility. However, the nature of the inputs and outputs and to some degree the processes can be generalized and relevant information can be brought together for educational programs.

As noted earlier, each step requires an input of subjective and objective information. While the specifics will vary from case to case, some aspects will remain constant in different contexts. For example, the scientific content involved in water problems is different from that of air but similar for other water problems, especially if the scientific information is disaggregated in categories such as lakes, rivers, groundwater, and oceans. This will remain relatively constant in any societal

FIGURE 1. MAJOR DECISION-MAKING STEPS IN ENVIRONMENTAL MANAGEMENT PROCESSES

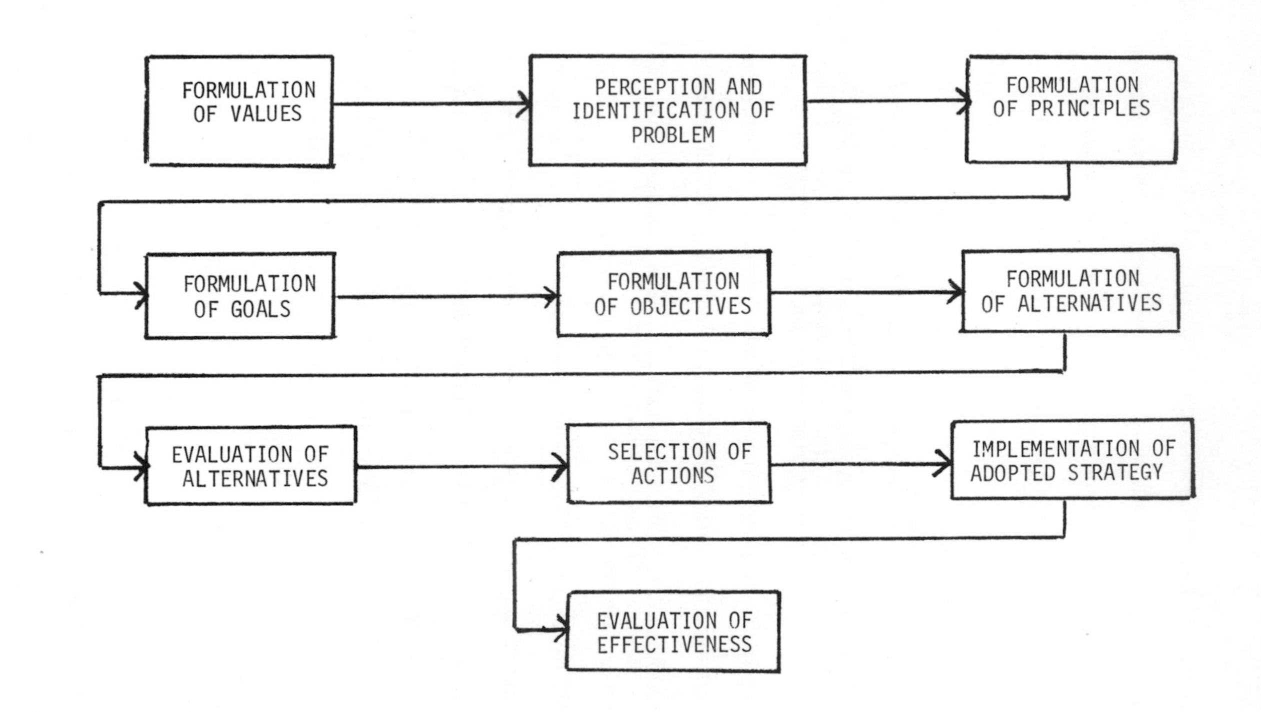

context. On the other hand societal values may be very different among cultures but within the same culture the values may be fairly constant despite the physical component of the environment one is considering.

The types of processes vary widely also. They may be corporate, executive, legislative, judicial, citizen, or other types of processes. It is possible, however, to generalize somewhat if one considers specific decision-making steps; for example, formulation of principles is generally a legislative function and choice of means to implement is generally an executive function. By considering specific actors or roles one can also develop some general contexts; for example, politicians play important and similar roles for some steps and a very small part in others.

Perhaps the greatest challenge of professional education in environmental management is to develop the generalizable contexts along knowledge, cultural, decision, and role categorizations. These must be broad enough to be useful for more than one manager in more than one societal context at more than one step in the process for more than one specific environmental problem. On the other hand, the contexts cannot be so broad that the amount of time and energy required to obtain the needed knowledge and insights would be prohibitive for any given person or that the coverage would necessarily be so superficial that it would be of little practical value to a manager.

The ultimate division of the major areas of substantive knowledge that we evolved in this project reflects our judgement of how this knowledge should initially be categorized. This is not final but it is based on the information and insights we were able to gather during the course of this project from persons in the academic and the managerial communities. This is discussed in more detail in the next section and presented in Part II.

In this project, we have only been able to take the most preliminary steps in determining the generalizable contexts along the lines of actors and the roles they play in specific decision-making steps. These determinations are very closely linked to the identification of the needs of the potential clients for educational programs and can thus emerge from the type of survey that is discussed in a subsequent section, "The Study of Actor/Role Interaction".

We have identified one area that can be developed and presented in depth to a wide range of "environmental managers" and still be relevant to much of what they are involved with. This is the area of analyzing and assessing environmental impacts. There have been many techniques developed and proposed which touch explicitly or implicitly on many of the questions of values, decision-making processes, and use of scientific knowledge that are so fundamental to environmental management. Some of the major issues and potentials of this area are discussed in the section "Environmental Assessment Techniques."

BODIES OF KNOWLEDGE

The primary emphasis of this project has been on the development of substantive outlines for the major bodies of knowledge that comprise the "field" of environmental management. These outlines have been supplemented with bibliographies containing sources where additional information on topics in the outline can be readily obtained. Many topics in the outlines are keyed to the relevant references. This material is in Part II.

We have divided the "field" of environmental management into the following subject areas:

Values and Perceptions

Ecology

Environmental Effects

Environmental Indicators

Environmental Impact Assessment Methodology

Modeling

Monitoring

Growth and Its Implications for the Future

Economics of Externalities

Environmental Law

Administrative Processes

Such a division is rather arbitrary and other sets of subject areas would also be useful. We have settled on these after much experimentation and discussion with others. We feel that it provides a reasonable compromise between a list of "academic" topics and a list of "operational" or "decision-oriented" topics. The former is important because we must rely heavily on the academic community for the knowledge in the various areas. The latter is important because we must make this information relevant to the practicing environmental manager.

We have in all cases focussed on functional knowledge--that which would be of use to a manager or planner in understanding the details or the context of some area relevant to his work. It is clear that to master these areas one would have to obtain a dozen Ph.D.'s and have many years of experience in a variety of areas. The challenge is to extract from these areas only the information that is directly useful for a specific function or role of environmental management.

This requires a fairly explicit set of criteria for selection of relevant material which in turn requires an extensive survey of the various roles that "environmental managers" have in the private and public sectors. This "market survey" of roles and functions is a key element in preparing educational programs. The C.E.I. has recently begun such a survey for multinational corporations. The results of the first exploratory effort are described in a later section in this Part.

The integration of the results of this M.I.T. project with the results of a complete "market survey" of the roles and functions environmental managers would be a matrix:

	Roles and Functions of Environmental Management
Outlines of Bodies of Knowledge	

Using the criteria developed in the market survey, the relevant topics in each outline of Part II could be selected for each role or function for which programs were to be designed. The references would then be helpful in locating the more detailed material needed for developing the programs.

In developing the outlines of Part II, we have gone through several iterations of literature review, interviews, and incorporation of the concepts discussed throughout this Part. Our objective has been to make each outline "academically sound", i.e., conforming to the general terminology and structure of the relevant academic areas, while presenting it in a manner of functional use to a manager. Thus, we have interviewed numerous academics and practitioners in developing the outlines and bibliographies.

The following sections contain a brief description of each subject area, the approach taken to it, and a general outline of the area. Part II contains these same descriptions and more detailed outlines.

Values and Perceptions

The knowledge of people's values and perceptions is a key to planning if the plan is to be supported by the people. Much emphasis has been placed on technological developments to increase our efficiency in using limited resources, but the question of whether such commitment of resources addresses our wants is often not asked. In order to formulate a plan which is to be responsive to people's present and future wants, the planner must identify their present values and inquire about how these values will change in the future. At the same time the planner must know what current problems are perceived by the people so that he can distribute resources in accordance to their priorities.

Every society exhibits a complex and shifting structure of values. This subject area presents an environmental manager with a spectrum of values which he may face in formulating a plan. If the plan is to address people's future needs, the environmental manager must surmise how present values will change in the future. This subject area assists the environmental manager to do so by providing information on the basis and development of values. The basic relationship between values and perception is also treated.

The primary purpose of this subject area is to draw the environmental manager's attention to the importance of knowing people's values and perceptions in planning and decision-making. It is doubtful that the subject area alone will enable him to formulate economically, institutionally, and politically feasible plans which will satisfy all wants, but together with other subject areas it will at least put him closer to that objective.

General Outline

Basis of Human Values

Factors in the Development of Values

Lists of Major Societal, Personal, and Institutional Values

Principal Variables that Shape Perception

Ecology

Ecology focuses on the interrelationships between living organisms and their environment. The recent heightened sensitivity of many people to the concept of environmental quality has assisted the development of ecology as an important branch of science recognized as relevant to everyday life. In order to comprehend the inherent complexities in this subject area, it is necessary to adopt a general and integrative approach. This approach is also necessary because ecology has evolved as an integrative science developing from a variety of disciplines including biology, geology, physiology, geography, meteorology, and many others. It is this interdisciplinary approach that has particular relevance to environmental management, as the study of ecology involves developing expertise for the purpose of evaluating the utilization of land, air and water habitats.

Since all human development activity takes place in some conjunction with natural systems, it is important to have a firm understanding of ecological limitations if our development is to be properly managed. Moreover, provision for the rational use of the earth's resource base implies consideration of the biological and physical precepts upon which all life functions. To fully appreciate the dynamic processes which may produce or act to disturb stability in the biosphere, it becomes essential to have a fundamental knowledge of underlying biological mechanisms.

Knowledge of these processes should assist a manager in asking penetrating questions about the ecological phenomena he confronts when performing environmental policy analysis. Some of these questions are: What basic ecological principles describe environmental dynamics? What are the relations between activities and effects on ecosystems? What kinds of responses can we expect from ecosystems when pollutants are introduced? How can planned developments be environmentally efficient? This subject area highlights consideration of these questions and many others.

General Outline

- Basic Principles and Concepts of Ecology
    - Structure
    - Energy
    - Biochemical Cycles
    - Growth and Productivity
    - Biological Transfers
    - Threshold Factors
    - Organization
    - Adaptations to Perturbations and Stresses
    - Time and Space
    - Resource Utilization
- Types of Ecosystems
    - Terrestrial
    - Freshwater
    - Marine
    - Estuarine

Environmental Effects

This subject area relates environmental effects to their causes. Associated with an activity is a set of fairly obvious direct or primary effects. However, the influence of the activity does not often terminate at these effects; instead, there is often a succession of derivative or secondary or indirect effects that follow one another like the links of a chain. Complicating the picture of the cause-and-effect relationship is the fact that an effect is generally not the result of one cause alone but of several converging causes. Also any effect may generate a number of direct and indirect effects which will occur at different times and in different places. An environmental manager should not infer from the structure of this subject area that cause and effect is a one-to-one relationship. A network is a better representation of the interaction and relationship among effects.

The purpose of this subject area is to expose an environmental manager to the complexity of the environment. However, even by broadening an environmental manager's perspectives on the spectrum of effects an activity may have on the environment, this subject area alone will not enable an environmental manager to contribute effectively to the preservation and enhancement of the environment, and at the same time achieve other objectives. After identifying all the probable effects of an activity, an environmental manager must be acquainted with what environmental indicators he can use to measure these effects, be aware of what values he should be concerned with, and be able to assess the total costs and benefits of the activity. Treatment of these subjects is provided in the subject areas of Environmental Indicators, Values and Perception, and Environmental Impact Assessment Methodology respectively. It must be emphasized that all subject areas are in some way related to one another, and the greatest use can be made out of each subject area only if the manager studies them all.

General Outline

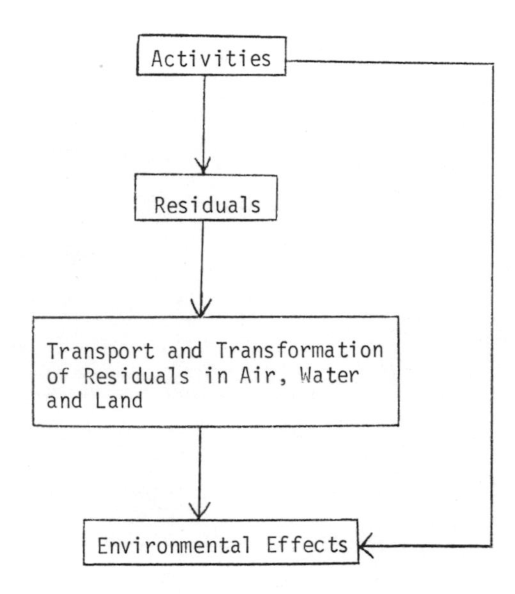

Environmental Indicators

As concern of environmental quality increases, there is an awareness growing among planners, decision-makers, and the public that more accurate and objective information on the status and trends of the environment is necessary to improve the formulation, implementation and communication of environmental policy. An environmental indicator, a number derived from a collection of statistics which quantitatively summarizes or measures the condition of the physical, ecological, social, economic or aesthetic environment, may often provide the kind of information required.

This subject area informs an environmental manager of the significant components in each environment where environmental indicators should be or have been formulated to reflect their trends and status. By providing information on who are the potential users of environmental indicators, how they can make use of environmental indicators, and what the limitations are on the use of indicators, the subject area enables the environmental manager to guide himself and others to use indicators discriminately. In some areas where indicators should be but have not been formulated, the subject area outlines the procedures to develop indicators and identifies what is needed in each procedure. Efforts in the development of environmental indicators are included in the last section of the outline. In all, the purpose of the subject area is to acquaint an environmental manager with a set of potentially effective tools for monitoring the status and trends of the environment and for evaluating the effectiveness of environmental policy.

The subject area is related to other subject areas and in particular Modeling and Monitoring. Relevant sections of other subject areas should be reviewed simultaneously in order to better understand and to make full use of this area.

General Outline

Needs of and Procedures for Developing Indicators

Types of Indicators

The Users, Uses, and Limitations of Indicators

Indicators Currently Used and Proposed

Environmental Impact Assessment Methodology

For decades the quantifiable costs and benefits, generally expressed in terms of dollars, of an activity have dominated the evaluation of the activity--particularly those that were undertaken to satisfy material wants. Values which could not be quantified or expressed in terms of dollars were only given superficial treatment and very little weight in decision-making. Environmental quality is one such value.

Many developed countries have had tremendous economic development but not without deterioration of their environment. The importance of values of environmental quality began to grow when the degradation of the environment became increasingly tangible. The awareness of environmental quality gradually pushed planners and decision-makers alike to change their concept of activity evaluation. Now emerging is a new concept that requires all probable impacts of an activity, quantifiable or unquantifiable, direct or indirect, to be evaluated in decision-making.

To make the new concept operative, an environmental manager must be able to systematically identify, predict, and evaluate all probable impacts of an activity. This subject area outlines the procedures to analyze objectively the impacts of an activity and to evaluate alternatives. These procedures can also help the environmental manager to be aware of what information he needs to select alternatives. Making an objective analysis of all probable impacts and evaluating alternatives in this sense are no easy tasks. This subject area provides the environmental manager insight on what obstacles may limit either process.

Although techniques to overcome these obstacles may not be available at present, the environmental manager should at least be aware of where the gaps are and focus his efforts on those gaps.

This subject area is closely related to the subject area of Environmental Effects which broadens the environmental manager's perspectives on the spectrum of effects which an activity may have on the environment, and the subject area of Environmental Indicators which acquaints him with a set of potentially effective tools. Together they augment the environmental manager's ability to assess the environmental impacts of an activity and to select the best choice.

General Outline

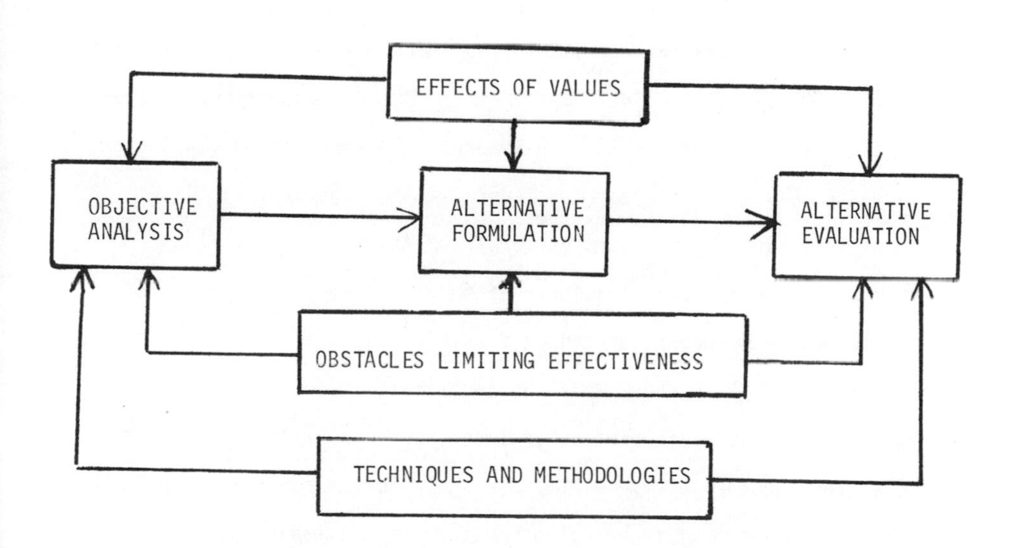

Modeling

The process of modeling environmental systems in some form has always been an important aid to environmental management. In the past, mental models were often cited directly under the claim of professional experience in order to justify certain particular assessments or policies. Recently, the technical state-of-the-art has developed rapidly and now includes extensive and complex dynamic programming techniques applied to simulate various environmental parameters on a large scale. The concept of modeling capability as an important tool that assists decision-making is thus quite valid. However, considerable difficulties can arise regarding how one gets specifically involved in the choice of models for a given problem. The well-informed environmental manager should have before him a comprehensive and previously studied list of criteria relating to the general appropriateness of each existing type of modeling system. He should also be well aware of a number of operational criteria that relate to model performance in general. The approach given in the outline for this subject area is a first step towards providing a grasp of these issues in a straightforward manner.

Given the recognized usefulness of modeling, it should be understood that some modeling techniques can be flexible enough to be used at almost any level of action. Modeling thus ties in with other subject areas to be considered in professional environmental management. For example, the tandem use of modeling for analysis and monitoring for verification of various technical parameters is one important way to determine the effectiveness of a given administrative control procedure.

The role of modeling is that of a technical tool to be used in conjunction with staff assistants and advisors to the decision-maker. Only when those directly involved with decision-making feel comfortable with these techniques will they really be used efficiently and purposefully. Thus, the approach of the outline for this subject area focuses on direct participation of the manager in the choice and use of various environmental models.

General Outline

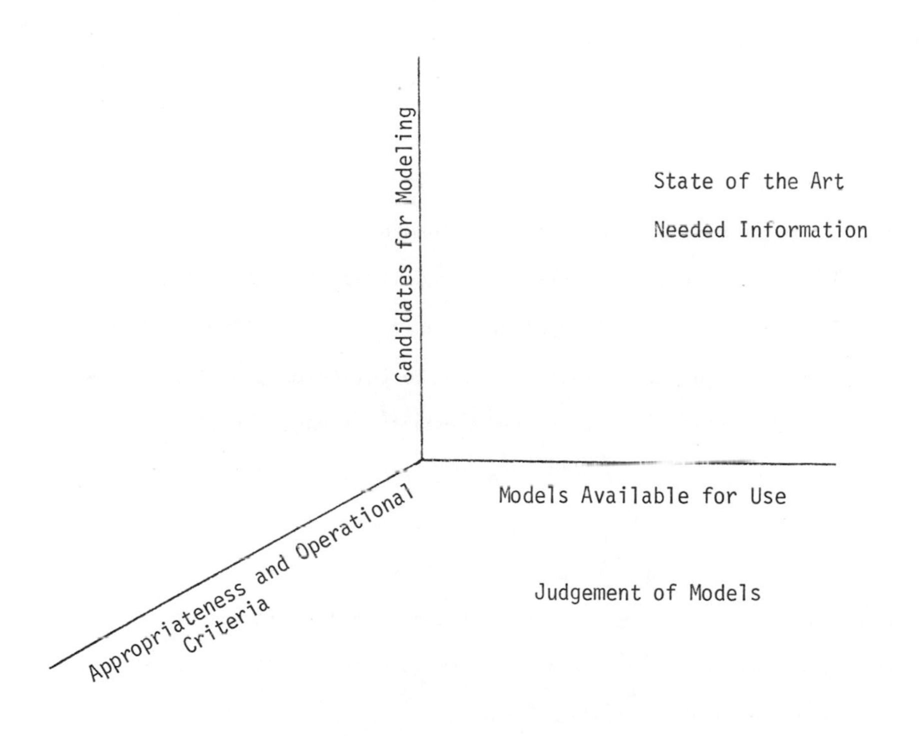

Monitoring

A fundamental responsibility of proper environmental management is the systematic accumulation of information about significant physical and social system parameters. Adequate monitoring capability can respond to a manager's need for data by providing comprehensive facts in a prompt manner within a tolerable degree of accuracy. Moreover, linkages among various monitoring systems allow decision-making on a broad level of action to be executed with the confidence that all critical segments of a given problem have been analyzed and reported.

Results from monitoring systems can provide the key factor in many problem assessment areas of environmental management, from initial problem definition (e.g. the carrying capacity of a given land area for extensive industrial development) to continuing post-project analysis (e.g. recording the actual migration of population to a given area after development has been initiated). Knowledge of the manner in which such information can be collected, and the limitations of such efforts, gives the environmental manager an initial basis for judgement as to the overall feasibility of assessment for any given undertaking. Another dimension of the utility of monitoring systems is related to the role these systems perform as part of an overall regulatory mechanism. The existence of certain monitoring techniques and systems may be the principal motivation for the enactment of standard-setting legislation.

It is apparent that monitoring processes can serve a crucial role in providing adequate assessment capability for a given issue. The outline presented here for this subject area is a first step towards providing the environmental manager with both an understanding of the various available systems and a set of criteria for use in choosing a system within a given situation. With these concepts in hand it is felt that the integration of monitoring efforts with other related tasks will be performed with a better measure of efficiency.

General Outline

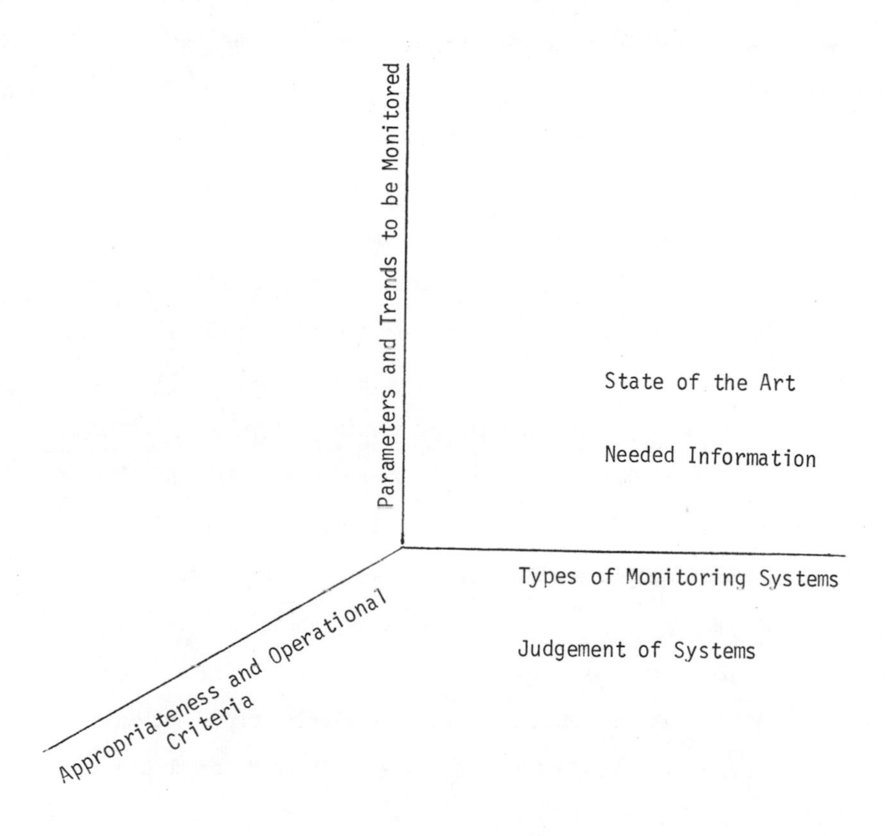

Growth and its Implications for the Future

Development of a comprehensive approach to environmental management concepts in an educational format requires some exposure to the underlying perspectives of broad-scale social behavior and trends. In the area of environmental affairs, the processes with which we are most interested are those that describe the past and present physical development modes of society and the various social motivations for these modes. This is because physical development modes provide the setting within which values are established and environmental disruptions result, and are perceived.

The most common mode of physical development by nation-states in the world today is a growth mode. This is characterized principally by attempts to expand the national industrial and economic base. Population growth often naturally accompanies this economic growth, and can fuel a need for further growth. The classical rationale for such expansion is that the quality of life increases as growth continues.

But today the traditional industrial growth mode is being disturbed by disruptive "side-effects" and also being attacked philosophically as not providing an adequate raison d'etre for daily life. Principal among these "side-effects" are the environmental implications of further growth and we must understand these in order to manage. At the same time, developing countries must be provided with the means for development consistent with an actual and not falsely perceived improvement in the quality of life. Understanding the benefits and hazards of growth is the first step in plotting viable futures for society.

General Outline

Basic Parameters Related to Growth

Benefits and Hazards of Growth

Evaluation of Adjustment Mechanisms

Public Policy Issues

Growth Parameters and Societal Indicators

Economics of Externalities

The issues and problems of environmental management can be very diverse, ranging from almost purely abstract concerns such as how to define the beauty of a parkland to extremely sophisticated analytical considerations such as air transport models for chemical particulate matter. However, in most circumstances the environmental manager will be confronted with problems that have significant dimensions in the economic sphere, since almost any proposal for environmental policy analysis requires the considerations of economic and technical implications.

An understanding of the fundamentals of economic systems is essential for the environmental manager and that knowledge is a prerequisite to the information in this subject area. What is often needed however is to focus that fundamental knowledge within a framework of economic principles related to classical theory, the theory of welfare economics, and the various interpretations of the causes of environmental "externalities" as they are explained in economic theory.

As we developed various concepts on the basis of the underlying economic theories available, we found that the most efficient means of packaging this information would be in the form of a discussion of the types of environmental "externalities" produced and how the economic impacts of these residuals could be forecast. After this, the principal issue area to consider is how economic costs and benefits can be assessed in analyzing the role of environmental pollution and depletion of resources. Certain specific questions of an economic nature arise in the private marketplace and these must be treated as well. Finally, much of the information discussed in this subject area will serve us well in attempts to elaborate an environmental impact assessment methodology.

General Outline

Economic Perspectives and Principles

Technical Perspectives of Residuals Generation

Economic Analysis Procedures

Important Issues

Environmental Law

The emerging professional practice of environmental management requires an understanding of the historical perspective of legal norms and the nature of modern legal approaches to environmental issues and problems. With respect to protecting environmental values, old concepts of common law including nuisance, trespass, and negligence concepts are not sufficient today. Such doctrines were formulated during an expansionary ethos in Western society, and often reflect the view that resources are not exhaustible. The same reasoning could be applied to procedural law--used historically for settling disputes, it makes no provision for guaranteeing rights to a livable environment.

The reason for a subject area in environmental law focuses on an examination of the functional value of the law to meet societal needs. Law can be considered as a principal means for social regulation in attempting to formalize social values into a rational code of principles. Different legal systems have arisen as a result of modifying values and principles. Law usually not only defines substantive codes of conduct but also delineates the roles of individuals and groups in society who police the prevailing codes.

As a formal system of rules, laws can be enacted on many organizational levels, any or all of which may be in the manager's domain. The manager must be familiar with the substantive content of laws and the manner in which they specify administrative actions for enforcing precedent, setting standards, and regulating actions. In order to plan the efficient use of resouces in legal encounters it is necessary to examine the broader issues of how laws are formulated in any given political system and what implicit roles are available in seeking legal changes.

It is felt that this particular perspective will provide the environmental manager with a valuable background because it encompasses issues necessary for a comprehensive understanding of the workings of the legal structure in different organizational modes.

General Outline

Values as a Basis for Law

Different Legal Systems

Controlling Activities and Actions

Administrative Processes

Environmental management can be considered as a process of discovering, analyzing, and making decisions about issues and problems concerning man's impact on the world and its resources. Within the usual societal context these actions are ascribed to both the individual in his daily life and the social institutions which transform resources into products.

For environmental management it is necessary to study the functions of administrative organizations from the perspective of how environmental control policies and strategies are formulated. Such an approach fits well with the other subject areas that have been developed since it details a type of actor/role interaction in the policy area that can be supplemented by consideration of other management dimensions. For example, explicitly considering the types of administrative controls available is one way of establishing how the range of monitoring and modeling strategies that are available (taken from development of the monitoring and modeling subject areas) can be applied to various environmental problems. Also, the economic dimension of environmental control is emphasized through a discussion of the rationale for implementing various controls. The role of environmental indicators and questions as to the availability and accuracy of relevant data need also to be considered from the viewpoint of the effectiveness of controls.

In considering environmental policy, any organization is forced with differences in the perception of environmental problems, the degree of commitment to values, and the ordering of policy priorities. An understanding of the available alternatives for control fills a needed gap in determining the options open to managerial action.

General Outline

Activities vs. Effects

Controls vs. Effectiveness

Organizations vs. Roles

124036

THE STUDY OF ACTOR/ROLE INTERACTIONS

In any environmental management process there are many actors involved. Each actor may also play many different roles throughout the process depending on the particular decision-making step of Figure 1 and on many other factors such as the authority one has at any given time or among any given set of other actors.

Thus, in trying to identify the roles and functions of "environmental managers", we are asking for much more information than the identification of a particular person in a particular position such as the chief engineer for water quality problems in an international organization. What one must know to develop meaningful education programs relevant to persons in many positions are the types of roles these people play in their professional interactions. For example, we need to know if he spends significant amounts of his time as a researcher, data-gatherer, data-interpreter, advisor to management, advisor to governments, public relations handler, trouble-shooter, problem definer, problem assessor, expert witness, resource allocator, dispute arbitrator, negotiator, and so on.

It is by analyzing the roles of "environmental managers" rather than their positions in organizational charts that we will ultimately be able to define what they need in order to do their jobs better. Furthermore, by analyzing all of the major roles in the environmental management process--of the concerned citizen, industrialist, lobbyist, bureaucrat, politician, judge, etc.--we learn about the actors with whom the environmental manager must interact and we learn about the nature of the interactions. This is obviously something the manager must know to be effective and there are very few ways he can learn this except through experience.

The number of possible roles for any one manager can be rather large. Every person who undertakes responsibility as a manager in this area is involved in an extensive network of official and unofficial encounters. Learning to accomodate these encounters and utilize them to further one's mission is a process accomplished more efficiently as more experience is obtained on the job. In all cases, a certain set of questions are asked and answered continuously and implicitly by those who find themselves in these positions: What roles actually exist as a function of position? What roles can I create or dissolve by choice and what roles am I obligated to play? What are the different possible interactions that I can have with other actors? How can I influence their performance to achieve results that I consider important? If conflicts occur between or among various actors and roles, how can they best be resolved? What are the steps by which action in support of policy can be galvanized? How can the use and manipulation of multiple roles help the advocacy of various policies in environmental issues?

These questions can form the basis for follow-up work that would build on the progress made during the course of this project. One of the principal recommendations which is made in the last section of this Part pertains to the kind of investigative effort needed to establish more accurately a classification system for environmental managers on the basis of their responsibility and function. Use of a questionnaire format that focuses on the above inquiries could prove to be essential.

For example, one of the first types of output to be expected from such a survey would be an extensive compilation of the types of actors who occupy what could be considered "official" classifications within their respective organizations. The basis for this type of output, of course, would have to be an examination of the nature of the various organizational modes that are open to the professional in a functional sense.

An initial classification of these organizational modes is not very difficult. One way to disaggregate organizations is via different levels of action: international, national, provincial, and municipal. Within each of these levels of action there are many types of subset classifications that could be made, the most obvious of which is probably governmental versus "quasi-governmental" versus non-governmental, where "quasi-governmental" would include agencies such as regional river basin commissions, metropolitan development commissions, and quite possibly government subsidized public enterprises. Non-governmental organizations include private industry, political action lobbying groups, the academic community, and perhaps the private citizen as an independent unit.

Further differentiation for an initial market survey should probably be made along functional lines once these initial classifications have been made. The conceptual framework of Figure 1 would be very useful for this purpose. As one focuses on decision steps it would also be necessary to refine the organizational classifications further, for example, in a governmental classification, the traditional breakdown of legislative versus executive/administrative versus judicial branches would be useful.

The result of this survey would be a system of functional positions, moving from (a) level of action to (b) nature of public/private sector relationships to (c) nature of the decision steps involved to (d) specific functional positions. For example, one type of position for an environmental manager would be chief research engineer for water quality problems in an international organization. This position might be one that deals with international levels of action from a quasi-governmental sector standpoint, in the assessment mode, with responsibilities over certain aspects of certain program development and certain personnel, and such a position is also subordinate to other positions.

Once a market survey establishes a set of functional positions, using this or another classification system, a beginning can be made in determining the existence of specific roles in each of the classifications by interviewing the persons in those positions. At this point an extensive study is required and a real time commitment is essential to allow the survey team to acquire an understanding in some depth as to the various types of roles that exist. Since we have seen that there are a vast multitude of positions possible in any system of formal classification, it will be possible only to explore a (hopefully) representative sample of positions in order to determine the most dominant and significant roles in the environmental management process.

Since the suggested market survey was not part of this project and has yet to be done, it is not clear what kind of classification system can be developed for analyzing the existence of roles as a function of position. However, if we assume that some sort of role classification can be established, we can make some tentative suggestions as to the forms of actor/role interaction that would then need to be studied. The study of these forms of interaction will provide a valuable input to the assembly of relevant educational packages for all types of environmental managers.

Some basic questions to consider include these: How important is personal sympathy with a given issue as a motivation for role adoption? What importance does a given actor attach to building personal and professional contacts with other actors? What types of scientific knowledge are considered essential for certain roles before these roles can be adopted? How does direct official use of authority affect ability to transfer between roles? What kinds of tactics can be used by different roles to anticipate conflict? What are the information exchange processes between various roles during an interaction?

The nature of the international dimension of actor/role interaction is rather unique and deserves special mention. During the course of our work with the C.E.I., as discussed later in this section, the types of interaction which the manager in an international organization faces were revealed as being burdened with an additional layer of bureaucratic requirements and a distinct disharmony of value systems which arise whenever the crossing of a national boundary implies the crossing of a cultural boundary. International civil servants face the daily problem of trying to understand the demands of national policies on any given environmental issue and transforming these demands into a multinational agreement, or else collecting information on environmental problems of various countries without the aid of a common data base. Managers working for multinational corporations often do not have an adequate knowledge of the social and legal frameworks for many societies, and thus cannot adequately forecast the future implications of transnational investments. Non-governmental groups working for the "public good" need a better understanding of how national cultural values form the basis for environmental principles in a given society, before they can hope to establish a common body of international environmental "rights and privileges." The international sector is a very fertile area for the type of actor/role research which we have discussed above.

The subject area of actor/role interaction is a vital one for professional education in environmental management. It provides the professional with a picture of the setting in which he operates, both in terms of the actors and organizations that are interested in his problems, and the tactics he can and does use to deal with these people in whatever role he chooses or finds himself. Further study of the subject area of actor/role interaction should include study of relevant reference works. A listing of the works that were helpful during the course of the project can be found in the last section of Part II.

In order to properly develop this subject area for operational purposes, an extensive market survey effort must be made to examine in depth the types of actors, roles, and tactics available in the real world regarding environmental affairs issues. Special emphasis needs to be placed on the international dimension of possible interactions because of the inherent added complexities. During the course of this project Mr. Joseph C. Perkowski of M.I.T. and Dr. Michael G. Royston of C.E.I. conducted a series of interviews with executives in nine European multinational corporations and a staff member of the

European Economic Community to determine the nature of their functions and on-the-job perspectives regarding environmental management.

In essence, these interviews constituted the beginnings of a possible "market survey" of professional environmental management roles. Those most concerned with environmental coordination in the following companies were interviewed: Akzo, The Netherlands; British Oxygen, England; British Petroleum, England; Ciba-Geigy, Switzerland; Dutch State Mines, The Netherlands; Euroc, Sweden; Imperial Chemical Industries Europa, Belgium; Kockums, Sweden; Pilkington, England.

It is important to note that the number of "samples" in this "mini-survey" may not be large enough to allow any kinds of policy conclusions to be drawn regarding the relative importance of various bodies of knowledge to environmental management functions. However, such a "mini-survey" does have value in the sense that it allows an initial conceptual framework to be constructed that may lead to future assessment of proposed educational strategies.

Such a framework is shown in Table 1 in the form of a matrix, juxtaposing the substantive bodies of knowledge developed by the M.I.T. staff with examples of various identified professional environmental management functions. These functions are of a descriptive nature, based upon our subjective impressions of the types of jobs that were being performed by the people interviewed. There is no intention here to make this list comprehensive; however, it does serve to define many of the professional roles which would include environmental management responsibilities.

Since no extensive effort was made to scientifically prepare a survey technique, the matrix only displays our perceptions of the

TABLE 1.

MATRIX OF SUBJECTIVELY RECORDED EXPLORATORY MARKET SURVEY RESULTS

ENVIRONMENTAL MANAGEMENT FUNCTIONS → / BODIES OF KNOWLEDGE ↓	Corporate Planning	Product Development	Public Relations	Government Liaison	Intra-Industry Liaison	Trade Union Relations	Production Emission Monitoring	Economics Department	Env. Mgmt. Personnel Supervision	Legal Department	"Crisis Management"	Occupational Safety & Health*
Ecology		1		1			2			1	2	
Administrative Processes	2		1	2	2	1		1	2	1	2	
Environmental Law	1		1	2	1	1	1	1		2	2	
Monitoring	1		1	1	1		2			1	2	
Modeling	1			1	1		2			1	1	
Actor/Role Interaction	2		2	2	2	2			1	2	2	
Growth	2	1	1					1				
Economics	2	1		1	1	1		2		1		
Values and Perception	1	1	2	2	2	2		1		2	2	
Assessment Methodologies	2	2	1	2	1		1	2		1	1	
Environmental Effects	1	2	2	1	1	1	2	2		2	2	
Environmental Indicators	1	1	2				2	2		1	2	

*Generally, not considered as an environmental management function

Code: 1 = perceived need noted for this particular body of knowledge as it applies to particular environmental management function

2 = very strong perceived need identified

responses of professionals in each management function as to the value of various bodies of knowledge. An arbitrary index of value was established and used in the matrix. Some useful overall impressions can be gathered from this display. For example, the function of "crisis management" was implicitly defined as that set of responsibilities which arose whenever an unusual incident involving environmental affairs arose quickly and needed to be resolved quickly. It seemed that this function requires most familiarity with many different bodies of knowledge, more so than any other function. On the other hand, personnel supervision and trade union relations seemed to be functions which have only marginal connection with professional environmental management as we have defined it, although some of the people interviewed stressed this aspect of their work as environmental co-ordinators. Occupational safety and health was generally considered to be in a different category of "internal" (to the company) environmental matters.

From the perspective of the "bodies of knowledge," much value seems to be placed in the knowledge and categorization of the subject of environmental effects. By contrast, the subject area of "growth and its implications for the future" was not considered as having high priority in the context of environmental management education. It is quite likely that the nature of the persons interviewed is such that short-term needs are paramount in their perception of their responsibilities for corporate sustainability.

It is also interesting to note one other aspect of the "mini-survey" that was relevant. During the interviews we asked for a listing of the perceived necessary qualities for a "generalist environmental manager".

The character description built from the responses had four principal criteria: (i) a basic level of general technical education; (ii) a substantial amount of production management experience; (iii) the ability to reason quickly, clearly, and systematically; and (iv) the personal quality of a high threshold of tact in interpersonal relations. Hopefully future surveys can further develop and refine this list.

The results of this "mini-survey," compiled mainly as a collection of substantive impressions, assisted us substantially in our development of lecture materials for courses at M.I.T. and C.E.I. However, it was increasingly evident that this type of information is needed in vastly more sophisticated form requiring substantial investment in research effort in order to build up an adequate basis for development of a professional education program in environmental management. Some major questions that arose include the following:

(1) To what extent is environmental management simply another facet of overall management practice, and to what extent is it uniquely different?

(2) Are the inherent uncertainties in environmental management issues actually much smaller in absolute magnitude than the uncertainties in other aspects of management decisions (e.g. prices, markets)?

(3) What is the proper personnel mix between central office co-ordination and environmental responsibility on a divisional level for a large organization that is multinational and highly diversified? This is a crucial issue whose importance was implied by the diversity of the responsibilities held by the people interviewed for the "mini-market-survey." We received the impression from our interviews that

the position of environmental management is still largely ill-defined in many organizations.

(4) If decentralized administrative control over environmental policy is implemented, how can we achieve balance between national environmental quality and regional environmental quality?

(5) How can we develop better research methods to determine legal trends in environmental control within many different countries?

(6) What are the advantages and disadvantages in administering effluent charges versus administering effluent standards as pollution control policy?

(7) What aspects of environmental management are different when considering the establishment of new facilities as opposed to the continued operation of established facilities?

In addition to these substantive questions, there are also questions involving the structure of meaningful educational experiences:

(1) Is it possible to effectively educate, in the same classroom situation, managers of both senior and junior levels? Environmental co-ordinators are often called upon to deal with other co-ordinators or with senior representatives of governments or political action groups, and thus it is possible that a mix of senior and junior people could not participate effectively since they would not share similar levels of experience.

(2) Participants are experienced managers and perceive significant potential benefits from an opportunity to share past experiences and present difficulties. How can this perceived need be incorporated efficiently into an educational seminar?

(3) What is the proper mix of participants in terms of organizational representation? How many representatives from governmental agencies should be invited? What kind of representation should there be from trade unions?

(4) How useful would it be to have prepared and distributed case studies for a seminar? What types of preparation are most valuable?

At present, there is no way to determine in any systematic fashion the various functions that comprise professional environmental management other than an extensive market survey. Without this kind of knowledge we cannot be certain that the educational programs we construct really reach the full spectrum of professional management needs in this area. Specifically, this means that the vertical elements of the matrix in Table 1 that describe environmental management functions have to be expanded and disaggregated. A market survey would develop functional classifications and then develop a set of educational needs for each function. This process is iterative in nature in that as management functions become more fully defined, the types of education required to fulfill the responsibilities exercised by those functions become more apparent. Then as progress is made in the development of educational materials, further questions can be developed on the nature of the user of those materials and answers to these questions can be sought.

ENVIRONMENTAL ASSESSMENT TECHNIQUES

Rationale and Nature of Examination

Today many if not all societies are acutely aware of the importance of the restoration and preservation of a healthful and enjoyable environment so that man can sustain himself into the future with an acceptable quality of life. It is obvious that the state of the future environment depends heavily on how both the public and the private sectors plan their societal activities. As environmental quality is gaining weight in decision-making there is an urgent need of a systematic procedure which would enable us to incorporate the consideration of environmental quality into and throughout the planning process.

Various approaches to environment impact assessment in the planning process of government agencies have been attempted and others are described in the literature. Examination of these approaches leads to some important observations about the value of assessment in general and the nature of the overall process of "impact assessment". The type of assessment methodology that is broadly relevant to problems of all kinds should be general enough to teach and at the same time specific enough to apply in realistic cases. Techniques that would be used should combine the subjective and objective aspects of environmental management in a form that allows the decision-maker to be aware of both aspects in full and assists him in expressing his own value-laden preferences. These techniques should provide a basic synthesizing function for the major bodies of knowledge in environmental management that are treated in Part II.

At the same time any technique should be applicable to any specific impact or set of subimpacts which have crucial significance. It should also be possible to teach such techniques to a manager in a fairly high-level position in order to enable him to better utilize the various specialists on his staff when confronting a complex environmental problem requiring a significant impact statement to be performed.

There are many methodologies now suggested for use and these methodologies can be compared to each other via a common set of explicit criteria. Rather than trying to define an _ad hoc_ set of evaluative criteria based only on general principles of environmental management and using this _ad hoc_ list to develop the "ideal" methodology, we have adopted an approach that stresses the potential advantages and disadvantages of current techniques.

Each of the following criteria were developed from an analysis of several available impact methodologies. Some criteria specifically relate to the need for identification of certain elements within an assessment methodology. Others relate to the evaluation process itself or to procedural elements such as information dissemination of the results of an assessment. Taken together, they constitute a critical checklist of ways in which impact assessment methodologies should be analyzed, and thus comprise important elements to consider in professional environmental education.

1. _Identification of the actors and values to be affected by the proposed action_. Although much has been said about the importance of objective analysis in environmental impact assessment, one must

realize that the entire process is subjective, i.e. basically value-oriented. Perceptions of environmental impacts are most meaningful when they can be interpreted in terms of social values. Without the knowledge of people's values, there is simply no basis for evaluation.

If people's values are used as the basis for evaluation, the assessor must identify what people like or dislike. One way to do this is to identify the parties that will be affected and to encourage them to actively participate in the assessment process. One of the purposes of encouraging public participation is to alert the assessor to what mitigating measures should be applied to a proposed development in order to minimize undesirable effects. Needless to say, not all values will be affected beneficially.

2. Identification of the activities that will be involved in development of a proposed project. A project development normally consists of a number of activities that will be carried out simultaneously or sequentially. For example, the proposal to produce more nuclear energy involves the mining of fuel, the actual construction of powerplants and other facilities in the nuclear fuel cycle, the transport of fuel and other materials, the operation and maintenance of powerplants and other facilities, and so on. The number of activities that will be involved depends upon the nature of the project. Prior to the prediction and evaluation of a composite or aggregate project impact, any assessment methodology must assist the analyst in identifying the significant activities that will cause the specific impacts.

3. Delineation of the major phases of a project. The chance of overlooking any activity that will have significant impact on the environment is minimized when the analyst can distinguish the major phases of the proposed project. The identification of impacts becomes easier because the process is more systematized. Another advantage of having a project divided into several major phases is that it provides the analyst with insight on why certain activities will be involved, and on where and when such activities will occur. The assessor can then make use of such information to investigate the cumulative impact of the activities which will occur either at different locations or at different times.

There are several ways to divide a project into major phases. For example, one can divide any powerplant project into the phases of site preparation, construction, operation and maintenance, and decommissioning. Other equally appropriate classifications could be made for other projects. An environmental impact assessment methodology should provide guidelines for choosing the best way to divide a project into appropriate phases.

4. Identification of impacts on all significant environmental parameters. Two fundamental processes in environmental impact assessment are to determine what does and what does not constitute "the environment" and "environmental impacts". In most societies environmental awareness stemmed from a few tangible phenomena such as air pollution, water pollution, landscape erosion, impairment of scenery, and so on. In the United States, in the implementation of the National Environmental Policy Act, these definitions have been enlarged to take into account almost all the elements of the human environment and all changes therein.

Environment is consequently construed to mean all interdependent living and non-living parts which make up man's ecosystem, and environmental impact refers to changes or conversions that occur. Accepting these definitions, the analyst will have to identify, predict, and evaluate a wide spectrum of impacts whose boundaries are probably not fully understandable at the beginning of the assessment process. Constraints on time and budget always require the analyst to make the best use of his limited resources. Therefore he must not waste his time on insignificant impacts. An environmental impact assessment methodology must assist the analyst to systematically identify impacts on all significant environmental components.

5. Identification of derived impacts resulting from the interaction of primary impacts. Associated with any proposed activity is a list of fairly obvious direct or primary impacts. However, there is always a succession of derivative, secondary, or indirect impacts which follow one another like the links of a chain. Some secondary impacts of a proposed highway development, for example, might be additional homes in the vicinity of the highway, commercial development in the highway vicinity that might lead to further ecological pressure on surrounding land, shift of modes of transportation in the region, and so on. These impacts are difficult to predict because the exact causal relationship of these impacts are not well perceived. Complicating the task is the fact that an impact is generally not the result of one cause alone but the result of several converging causes. Also any single impact may generate a number of impacts which occur at different times. An environmental impact assessment methodology must assist the analyst in predicting and evaluating the impacts resulting from the interaction among impacts such that all true benefits and costs of the proposed project can be adequately evaluated.

6. Identification of meaningful parameters which indicate the status and trends of significant environmental components. Environmental concerns can be arbitrarily classified as concerns for the physical, ecological, social, economic, aesthetic, and cultural environments. Within each of these categories there are a number of components which are of public concern and worthy of separate consideration. A sample list of components in the social environment includes national security, health, education, safety, opportunity, economic growth, transportation, and various amenities. Impacts on these components are evaluated in terms of the social values attached to them.

An additional point to mention is that not all environmental parameters are equally important. Under certain circumstances it is more important to have more information on some parameters relative to others. Constrained by time and budget, the analyst must be able to identify the most important parameters for a given assessment and spend a larger part of his effort in acquiring information about them. An environmental impact assessment methodology should inform the analyst about the units of measure used to express an environmental parameter and also indicate which components have no such parameter available.

7. Description of the uncertainty of known impacts. Often pervading environmental impact assessment is the question of uncertainty, and in some instances it is the determining factor in a decision. The most common problem of uncertainty is that associated with prediction. There is simply no way to predict exactly how each significant environmental component will be affected by a proposed project. For example, the deleterious effect of impoundment on a biological system can at best only be surmised, because of the understanding of the structure of a biological system and the flow of energy and mass within it is imperfect.

The crucial question now is how to select the best alternative in the midst of uncertainty. Economists and systems analysts deal with this problem in their accounting of expected costs and benefits. The assessment of environmental benefits and costs is more complicated than common cost and benefit analysis because not all environmental benefits and costs can be expressed in terms of a single measure, and the importance of accounting for uncertainty is thus enhanced.

8. Description of the dynamics of significant environmental impacts. An important aspect worthy of consideration in environmental impact assessment is the dynamics of significant environmental impacts. The dynamics can best be described by answering the following questions: when will the impact occur? will the impact change over time? how long will the impact last?

Knowing the dynamics of the impact on an environmental component is a prerequisite to investigating whether and when such impact will be severe enough so that it triggers other, unexpected impacts. It is doubtful that the present worth concept in economic analysis can be used to derive a measure which represents the aggregate effect of an impact over time. Nevertheless, if the importance and magnitude of all significant environmental impacts are to be considered in the assessment process, the assessment methodology must provide a means to enable the analyst to account for impact dynamics.

9. Accountancy for the incidence of environmental costs and benefits. Most analyses use overall net economic benefit or overall cost/benefit ratios as the basis for alternative selection when assessing environmental programs. The incidence or the distribution of benefits and costs have not been given much weight in the analytical process of selection. Public

projects, however, will almost always affect some groups or interests beneficially and others adversely. Formal methodologies for including the distributional aspects of impact assessment do not really exist at this time.

Unfortunately some studies have indicated that because of overlooking the incidence of costs and benefits in selecting among alternatives for project proposals in the past, the quality of the environment, particularly of common media such as air and water, available to an individual may be proportional principally to his socioeconomic status. This point of view needs to be pursued further, especially by serious consideration of the incidence and distribution of environmental costs and benefits. If improvement of environmental quality for all is an objective of public projects, then an impact assessment methodology must assist the analyst in accounting for the incidence of all costs and benefits in the selection of alternatives.

10. Determination of the significance of impacts. Changes or conversions of any environmental component are relatively useless unless such changes or conversions can be interpreted in terms of social value. Quantitative measures which indicate how an environmental component is affected are desirable because they can be used to compare objectively how each alternative affects an environmental component. While the relation of quantitative impacts to social value structures may always be explicitly difficult, an environmental impact assessment methodology must always have some sort of procedure for calculating the quantitative impacts that are significant for any given proposed project.

11. Comparison of short-term and long-term costs and benefits. If the environment includes all interdependent living and non-living parts which make up man's ecosystem, the environmental impact analyst must determine the relationship between local short-term uses of man's total surroundings and the maintenance and enhancement of total long-term productivity. Almost all project developments require commitment of some sort of resources. Commitment of an irretrievable resource is final if the resource committed will never be available again. An example of this point is the consumptive use of fossil fuels. The analyst must examine the retrievability of resources that will be committed, the range of potential uses of such resources, and the nature of the committment.

Economic techniques such as discount rate analysis may not be very helpful in the analysis of intangible environmental factors with long-term implications such as for example the genetic resources of a given plant species. Notwithstanding the difficulty of evaluating short-term and long-term costs and benefits, however, the analyst must regard as a cost any future option foreclosed as a result of the recommended action. An environmental impact assessment methodology must assist the analyst in comparing the short-term and long-term costs and benefits of all feasible alternatives such that he can select the alternative which will allow the greatest range of future uses of the environment.

12. Facilitating the comparison of alternatives. There is a definite need for a technique to effectively compare alternatives. If the result of an environmental impact assessment is simply a display of the impacts of all alternatives, some helpful work has been done. However, if in

addition, there are mechanisms of comparison available to judge the relative importance of certain impacts that can be compared, the analyst has moved the decision-maker one step further.

This can be a difficult job for the analyst. He will definitely have to consider the uncertainty with which a given impact of a given alternative will occur, when and where it will occur, and who will be affected. An intelligent choice among alternatives requires something more than mere intuitive judgement. An impact assessment methodology that points the way towards intercomparison of alternatives in at least some of their relative aspects will satisfy at least partially a very important evaluative criteria.

13. Dissemination of the results of the assessment. One of the primary purposes of an environmental impact statement (as it is commonly developed in the United States) is to provide information to the public on impacts of possible alternatives for a given project. However the objective of informing the public is often hampered by a few elements which are common among many impact statements. Some statements overwhelm the reviewer with vast amounts of poorly organized, remotely related, and often repetitious information. Review of such statements has been a frustrating experience; intelligent critique is almost impossible. Other impact statements are simply so brief that there is no basis for discussion. In addition, the techniques and implicit assumptions used to predict environmental impacts are rarely explicitly mentioned in the statement. Consequently evaluation of technical adequacy is, if not absolutely impossible, time-consuming and painstaking.

In light of the potential of an environmental impact statement an environmental impact assessment methodology must provide a format to translate the results of assessment into lucid and useful terms for review and use by the public and private parties. Specifically, the results of assessment must be organized and presented in a manner which addresses the following points:

a. Intelligible to reviewers of different degree of sophistication;

b. Enable the reviewer to check whether all activities having significant environmental impact have been studied;

c. Enable the reviewer to check whether impacts on all significant environmental components have been assessed;

d. Enable the reviewer to examine the technical adequacy of the assessment;

e. Enable the reviewer to follow how impacts are assessed; and

f. Enable the reviewer to follow how alternatives are selected.

In summarizing the discussion of criteria with regards to the general applicability of environmental impact assessment methodologies, some overall criteria can be considered. A methodology must be comprehensive in its framework for impact assessment, along the dimensions of time, space, and nature of affected actors. The framework used to record impacts must be orderly and systematic so as to permit replication from project to project. Any methodology must be practical in the sense that by using it an assessment can be completed in a short period of time without an unreasonably substantial commitment of manpower. Also, a methodology must be applicable not only to a certain type of project but to a wide range of programs that include many different projects having many varied impacts related to the environment. To satisfy this point, the approach used must be multidisciplinary and therefore the analyst must be prepared to consult with many different specialists and integrate their results into a coherent framework on display.

After thorough review of the methodologies studied and after consideration in depth of the criteria detailed above, it has become apparent that central to the whole process of assessment is the consideration of values. Somehow the analyst must always keep uppermost in his mind the range of values that will come into focus when the proposed project is analyzed. Some of these values will be relatively objective in nature, such as the magnitude of net resource use of a given project. Others will be highly subjective, such as the perception of land use degradation as the result of a project. Consideration of the value-laden basis of assessment leads to the conclusion that effective monitoring programs for determining post-project changes in value structure and the overall status of environmental parameters can play a very important role.

Finally, it is worth noting that no assessment methodology either presently used or seriously proposed for the near-term future really approaches the ideal format for assessment in the abstract. Parallel development of techniques for long-range planning in various environmental sectors such as energy and land use is becoming more and more of a vital element for today's society. Interactions between and among these sectors are at present poorly known, both subjectively and in regard to objective facts. In the ideal case, the ultimate assessment procedure would be a display of the interactions among all important social sectors (i.e. employment, land use, pollution, etc.) that would result for each given alternative for a proposed program. Then society, if presented at large with these data, could make a choice based on its own perception of the relative worth of all its various values. Such an ideal procedure will never be reached, but having its image as a goal helps the analyst to function best within the limited techniques available today.

MAJOR CONCLUSIONS AND RECOMMENDATIONS

The primary thrust of this project has been the collection, compilation, organization, and presentation of outlines of the principal topics in the major areas of substantive knowledge in the "field" of environmental management. We feel that Part II represents the most complete and most useful form for these areas that can be assembled without further research on the numerous functions and roles of environmental management. This book should, however, be considered as an interim step in the development and synthesis of knowledge from numerous disciplines to provide the raw materials for developing educational programs. In this section we shall summarize some of the things that should be done in the next steps so that the operational goals of professional education in environmental management can be met.

Development and Synthesis of Major Areas of Substantive Knowledge

The information contained in the outlines of the bodies of knowledge of Part II developed during the project could be used in many different ways. In order to draw conclusions as to future directions for substantive research in professional environmental management, we should illustrate some ways in which our present information can be expanded.

One of the most important criteria for this type of information is that it make a contribution to the operational function of an environmental manager. This operational orientation appears very strongly in the discussion of environmental impact assessment methodology. As that discussion indicates, criteria for analyzing the value of any given impact assessment technique need to be expanded. Also, interdisciplinary collaboration on the development of standardized indicators of complex ecosystem impacts needs to be accelerated. From another perspective, it is apparent that additional work needs to be done in developing more rational means of incorporating an economic assessment of abstract values and expressions (such as stress, crowding, etc.) in order to more fully acknowledge the influence of these factors when one is considering alternative uses of resources.

The nature of fundamental personal and societal values is very poorly understood. But these values can contribute to the formulation of principles which are then given legal authority and appear as another sort of constraint on environmental management strategies. The manager of a corporate investment program or a government development agency must consider the actions and reactions of the people he interacts with. He must ask himself not only "What tactics are these people using that result in conflict with my strategies, and how should I respond?", but also "What is the underlying motivation for their use of these tactics?" Thus he should be interested in supporting research directed to understanding the evolution of conflict situations from basic value differences.

Research is also needed for systematically organizing prevalent legal concepts and regulatory control for environmental policy in many different societies. This conclusion and recommendation arose from our "mini-market-survey" and bears directly on the operational role of the corporate environmental manager. A much closer examination of the economic implications of different forms of fees and charges for pollution control also needs to be made. Such work could provide a more efficient and rational allocation of both fiscal revenues and the available carrying capacity of environmental media.

Modeling of physical and social systems provides a potentially powerful tool in understanding the managing of environmental resources and man's activities. The practical use of results from the modeling of environmental systems has always been limited by both the reluctance of professional specialists from different disciplines to collaborate in an interdisciplinary effort and also the difficulty of communicating results of technical analytic efforts in a form understandable by decision-makers. Long-term payoffs can result if efforts are made to reduce these obstacles.

Although the outlines that we have presented are probably as applicable to the educational needs of managers and planners in developing countries as in developed countries, the detailed information available for expanding the outlines in large part concentrates on the needs of developed industrialized nations. This is essentially a consequence of where academic and scholarly opportunities, priorities, and resources have been concentrated in the past. Thus, there is a great need to develop the material in these outlines along lines uniquely suitable for the needs of developing countries.

Some of this can be accomplished through a compilation of literature from developing countries that the project staff did not have ready access to in the course of our work. We suspect, however, that projects will have to be initiated to generate new knowledge about the environmental and management elements and processes of developing countries. We also feel that this could probably best be done in the context of the most pressing environmental problems of those countries and then broadened to the more general needs of environmental management education rather than taking the "academic" view of developing specialized areas and then integrating them.

Development of Further Market Survey Efforts

The initial efforts to identify the specific needs of environmental managers in developed countries and to gain a better understanding of actor/role interactions have produced some results. These have allowed us to design some experimental educational programs which in turn will yield more information about the needs of environmental managers.

It is essential that this work be continued if meaningful educational programs are to be developed. This report has outlined an approach to a major study of actor/role interactions and for a survey of the roles and functions of major actors in environmental management processes. The categorizations of roles and functions must be expanded on the basis of in-depth field interviews and a thorough review of the relevant literature of management administration. The level and nature of professional education required for such management classification could then be developed in greater detail.

This work would begin by establishing a list of government agencies and private sector organizations by type and by environmental issue. "Quasi-governmental" bodies and political lobby groups should also be included. For this list of organizations, a survey form should be completed indicating the various possible perceived roles by all the actors in each organization within an environmental management context.

Then a representative sample of professional people who would be likely to play these roles should be intensively interviewed. Their reactions to proposed programs and perceptions of educational needs will serve as the basis for further refinement of program development and an identification of the major substantive and process-oriented educational issues that environmental management must confront over the long- and short-term. An ongoing effort to establish and maintain lines of communication with these people is essential in the design of programs to meet their needs.

The Centre d'Etudes Industrielles (C.E.I.) is presently undertaking a program along these lines for multinational corporations primarily in Europe. These activities should probably be expanded by C.E.I. or other organizations to examine the roles and needs in other developed countries.

For developing countries, a somewhat different approach is probably desirable. One suggestion has been made several times by persons whom we have interviewed, particularly by some of the staff of the Environmental Management Institute of the University of Southern California who have several decades of international experience through the School of Public Administration. The approach requires the assembling of a small group of environmental management educators from several of the relevant disciplines composed of persons from developed and developing countries. This group would be sponsored by a body such as the United Nations Environmental Programme.

The group would conduct a series of exploratory seminar/discussion sessions in regions of the developing world. The persons attending would be those from the region with operational responsibility for environmental management and planning and those from the academic community who would be able to develop educational programs. One of the purposes of the sessions would be to present some of the major concepts of environmental management and describe which substantive areas were available for educational programs. The other purposes would be to explore which environmental problems in the region should be used as a focal point for initial educational efforts, what the managers would like to learn about these problems, and what resources were available in the region for designing and implementing new programs.

The group would then work with those in the region to prepare seminars, short courses, and discussion sessions on the problem areas. These would eventually be broadened to other areas of concern to the region and might ultimately result in programs with sufficient breadth to be useful for managers with a wide range of substantive and management interests and responsibilities.

To the best of our knowledge this type of effort has not yet been undertaken. We feel that if meaningful education programs are to be instituted in developing countries that this or a similar type of activity must be undertaken.

It is our hope that this book will provide a part of the foundation on which international programs of professional education for environmental management will eventually be established. In addition to the development of the materials in this book, we have also spent considerable effort in initiating and assisting processes that are essential to these long-term goals and which have the potential of continuously evolving these materials and of developing programs using them.

PART II

OUTLINES AND BIBLIOGRAPHIES FOR SUBJECT AREAS IN ENVIRONMENTAL MANAGEMENT

VALUES AND PERCEPTIONS

The knowledge of people's values and perceptions is a key to planning if the plan is to be supported by the people. Much emphasis has been placed on technological developments to increase our efficiency in using limited resources, but the question of whether such commitment of resources addresses our wants is often not asked. In order to formulate a plan which is to be responsive to people's present and future wants, the planner must identify their present values and inquire about how these values will change in the future. At the same time the planner must know what current problems are perceived by the people so that he can distribute resources in accordance to their priorities.

Every society exhibits a complex and shifting structure of values. This subject area presents an environmental manager with a spectrum of values which he may face in formulating a plan. If the plan is to address people's future needs, the environmental manager must surmise how present values will change in the future. This subject area assists the environmental manager to do so by providing information on the basis and development of values. The basic relationship between values and perception is also treated.

The primary purposes of this subject area is to draw the environmental manager's attention to the importance of knowing people's values and perceptions in planning and decision-making. It is doubtful that the subject area alone will enable him to formulate economically, institutionally, and politically feasible plans which will satisfy all wants, but together with other subject areas it will at least put him closer to that objective.

VALUES AND PERCEPTIONS

General Outline

Basis of Human Values

Factors in the Development of Values

Lists of Major Societal, Personal, and Institutional Values

Principal Variables that Shape Perception

VALUES AND PERCEPTIONS

Detailed Outline

I. Basic relationship between values and perception*[4,5,16,28,36,37,47]

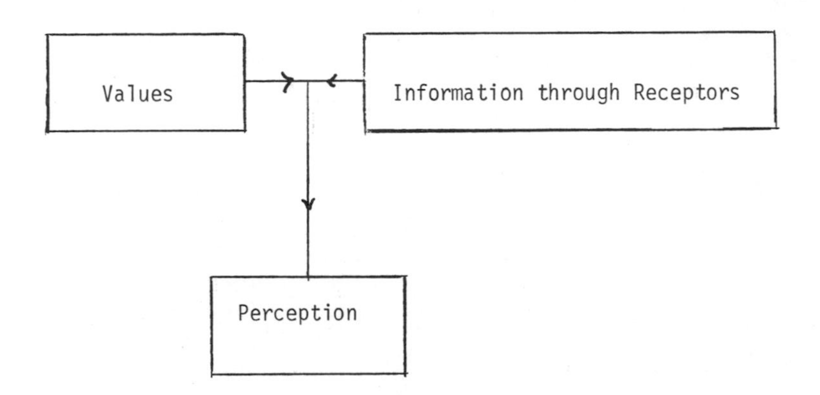

A. Definition - Perception is an image resulting of values with the information received by an individual, institute or society through the receptor.

B. Principal variables which shape perception

1. The existence of values per se
2. The ranking of values
3. The degree of dissatisfaction of values
4. The nature of information received

II. The basis of values

A. The needs for survival

1. The desire for physical needs
2. The desire for safety needs

B. The cognitive map theory

---

*Superscripted numbers throughout this Part refer to items in the bibliographies immediately following each section.

II. B. (con't)

1. Values vs. cognitive map related processes [2,14,29,30,31,42,44,54]

COGNITIVE MAP RELATED PROCESSES	VALUES
a. Recognition	The bias towards interpreting new events in familiar terms
b. Prediction	The enjoyment involved in predicting possible outcomes in uncertain circumstances
c. Evaluation	The delight in making distinction
d. Action	The delight to act in such a way to have predictable results

2. Socioeconomic aspects which affect the values resulting from the cognitive map related processes
   a. Age
   b. Sex
   c. Ethnic
   d. Education
   e. Occupation
   f. Income

C. Inherent human characteristics

1. Fascination with action, competition, sex
2. Irritated with confinement
3. Other characteristics e.g.- selfishness, greediness, etc.

III. The development of values [20,22]

A. The pace of development

1. Gradual

2. Abrupt

B. The areal scale of development

1. Local

2. Regional

3. National

4. Multi-national

5. Global

C. Values resulted from

1. Aggregation of values

a. Personal and groups of common values

b. Institutional/Professional

c. Societal

2. Historical development in

a. Religion

(1) Monotheism
(2) Pantheism
(3) Orthodoxy
(4) Judaism
(5) Hinduism
(6) Mormonism

b. Culture

(1) Agricultural
(2) Commercial
(3) Industrial
(4) Mixed

c. Economics

(1) Entrepreneur
(2) Family-type
(3) Corporation
(4) Giant enterprise
(5) Centrally-planned
(6) Market-planned

d. Social environment

(1) Mobility
(2) Education
(3) Technology - industrial revolution, automation
(4) Communication evolution
(5) Structure

e. Politics

(1) Democracy
(2) Communism
(3) Socialism

f. Physical environment

(1) Pollution
(2) Depletion of resources

g. Ecological environment

(1) Diversity
(2) Stability
(3) Resiliency

IV. Subjective Values [34,49,50,51]

A. Scope of Values

1. Local
2. Regional
3. National
4. Multi-national
5. Global

B. Types of Values

1. Personal
2. Societal
3. Professional/Institutional

C. Values [6]

1. Security
2. Health [39]

3. Economics

    a. Economic growth

    b. Net income

    c. Inter-regional trade

    d. Currency value and stability

4. Environmental quality 1,3,8,9,10,11,12,17,18,19,21,23,24,35, 39,40,41,43,45,46,53,55

5. Ecosystem

6. Social well-being

    a. Welfare

    b. Education

    c. Housing

    d. Service availability and quality

    e. Access mobility

    f. Community identity, cultural creed and political philosophy

    g. Recreation

    h. Amenities

    i. Employment

7. Aesthetics

V. The ranking of values

A. By free choice

1. With respect to the hierarchy of needs

2. Actor/role interaction

    a. Intentional influence

    b. Unintentional influence

    c. Dependence on the effectiveness of communication

3. Arbitrary

B. By external force

VI. Means to delineate perception [15,25,38,48,52]

A. Approaches

1. Question - response type

2. Models, drawings, and photographs - response type

3. Statistical analysis of response

B. Variables

1. Scales of approach

2. Choice of samples

3. Inconsistency of response

4. Accuracy of response

VALUES AND PERCEPTIONS

Bibliography

1. An Assessment of Noise Concern in Other Nations. Vol. I. Informatics, Inc., Bethesda, Md. Dec. 1971. 497 pp.

2. Anderson, Richard C., and David P. Ausubel. Readings in the Psychology of Cognition. N.Y.: Holt, Rinehart & Winston, 1965.

3. Appleyard, Donald, and Mark Lintell. The Env. Quality of City Streets: The Residents' Viewpoint. Inst. of Urban and Regional Development, U. of California, Berkeley. Reprint #77. Mar. 1972. 17 pp.

4. Beardslee, David C., and Michael Wertheimer. Readings in Perception. Princeton: Van Nostrand Co. 1958.

5. Berelson B., and G. Steiner. Human Behavior -- An Inventory of Scientific Findings. Harcourt, Brace & World. 1964.

6. Cantril, H. The Pattern of Human Concerns. New Brunswick, N.J.: Rutgers U. Press. 1966.

7. Clawson, Marion. Methods of Measuring the Demand for and Value of Outdoor Recreation. Reprint #10, RFF Inc., Wash., D.C.: Resources for the Future. 1959.

8. Coates, G.J., and K.M. Moffett (eds.) Response to Environment. Raleigh, N.C.: School of Design, N.C. State U. 1969.

9. Cudnohufsky, W.L. "Qualities of Landscapes that Please ," The Massachusetts Heritage Vol. 8, No. 4. December 1970.

10. Craik, Kenneth. "The Environmental Dispositions of Environmental Decision-Makers," Annals of the American Academy of Social & Political Science. May 1970. pp. 87-94.

11. Dasgupta, Satadal. Attitudes of Local Residents toward Watershed Development. Social Science Research Center Preliminary Report 18, Mississippi State U.: Water Resources Research Inst. May 1967.

12. Degroot, Ido. "Trends in Public Attitudes toward Air Pollution," J. of the Air Pollution Control Association. 17 October, 1967. pp. 679-81.

13. Dubos, Rene. Man Adapting: An Exploration into his Limitations and Potentialities. New Haven: Yale U. Press, 1965.

14. Environment and Behavior. Cognitive Representations of Man's Environment Vol. 2, No. 1. 1970.

15. Fishbein, Martin (ed.) Readings in Attitude Theory and Measurement. N.Y.: John Wiley & Sons, Ltd. 1967.

16. Gibson, James J. The Perception of the Visual World. Boston: Houghton Mifflin Co. 1950.

17. Hare, Charles T., and Karl J. Springer. Public Response to Diesel Engine Exhaust Odors. Dept. of Automotive Research, Southwest Research Inst., San Antonio, Tex. Apr. 1971. 89 pp.

18. Highway Research Board, Div. of Engineering. Transportation & Community Values. National Research Council and National Academy of Sciences, National Academy of Engineering, Wash., D.C. 1969.

19. Ibsen, Charles A. and John A. Ballwez. Public Perceptions of Water Resource Problems. Water Resources Research Center Bulletin 29, Blackburg, Virginia: Virginia Polytechnic Inst. Sept. 1969.

20. Insko, Chester A. Theories of Attitude Change. N.Y.: Appleton-Century-Crofts. 1967.

21. Jacoby, Lewis R. Perception of Noise, Air and Water Pollution in Detroit. Michigan Geographical Publication, U. of Michigan, Ann Arbor, Michigan.

22. Jahoda, Marie, and N. Warren. Attitudes: Selected Readings. "Theory and Method, Attitude Methodology." Baltimore: Penguin Books. pp. 287-353.

23. Jonsson, Erland. "Annoyance Reactions to External Environmental Factors in Different Sociological Groups." Acta Sociologica, 7. 1963. pp. 229-59.

24. Jonsson, Erland, et al. "Annoyance Reactions to Traffic Noise in Italy and Sweden." Archives of Environmental Health. 19 Nov. 1969. pp. 692-99.

25. Kaplan, Rachel. "Some Methods and Strategies in the Prediction of Preference." Landscape Assessment: Values, Perception, and Resources. E. H. Zube, J. G. Fabos and R. U. Brush (eds.). 1974.

26. Kaplan, Rachel. "The Dimensions of the Visual Environment: Methodological Considerations." Env. Design: Research & Practice. Proceedings of the Env. Design Research Association, Conference Three, Los Angeles, 1972: W. J. Mitchell, editor.

27. Kaplan, Steven. "Adaptation, Structure, & Knowledge: A Biological Perspective." Env. Design: Research & Practice. W. J. Mitchell, editor.

28. Kaplan, Stephen. "An Informal Model for the Prediction of Preference." Landscape Assessment: Values, Perception, and Resources. E. H. Zube, J. G. Fabos & R. O. Brush (eds.). 1974.

29. Kaplan, Stephen. "Cognitive Maps, Human Needs and the Designed Environment," Env. Design Research. Stroudsberg, Pa.: Dowden, Hutcheison & Ross. 1973. W.F.E. Preiser (ed.)

30. Kaplan, Stephen. "Cognitive Maps in Perception & Thought," Cognitive Mapping: Images of Spatial Env. Chicago, Ill.: Aldine. 1973. Roger M. Downs & David Stea (eds.).

31. Kaplan, Stephen. "The Challenge of Env. Psychology: A Proposal for a New Functionalism," American Psychologist. 1972. 27. 140-3.

32. Kates, Robert. "Human Perceptions of the Environment," Paper given at UNESCO Conference on Environment, Helsinki. 1970.

33. Kates, W., and Joachin F. Wohwill. "Man's Response to the Physical Environment," J. of Social Issues, XXII, No. 4. Oct. 1966.

34. Kaynor, E.R., Irving Howards. Attitudes, Values, & Perceptions in Water Resource Decision-Making within a Metropolitan Area. Publication #29, Water Resources Research Center, U. of Massachusetts Amherst.

35. Kornhauser, Arthur. Detroit as the People see it: A Survey of Attitudes in an Industrial City. Detroit: Wayne U. Press. 1952.

36. Lowenthal, David, et al. An Analysis of Environmental Perception. A Second interim report to Resources for the Future, Inc. American Geography Society, N.Y., & Harvard U., Cambridge, MA. Nov. 2, 1967.

37. Lowenthal, David (ed.) Environmental Perception & Behavior. Chicago: U. of Chicago, Dept. of Geography, Research Paper No. 109. 1967.

38. Measuring Attitudes Toward Water Use Priorities, Bulletin #50. Water Resources Research Center, Virginia Polytechnic Inst., Blacksburg, Virginia.

39. Medalia, Nahum Z. "Air Pollution as a Socio-environmental Health Problem: A Survey Report," J. of Health and Human Behavior, 5 (1964). pp. 154-65.

40. Medalia, Nahum Z., and A.L. Finkner. Community Perception of Air Quality: An Opinion Survey in Clarkston, Washington. Robert A. Taft Sanitary Engineering Center, Cincinnati, Ohio. June 1965. 115 pp. PB-168 875.

41. Mills, C.H., and D.W. Robinson. "The Subjective Rating of Motor Vehicle Noise," Noise: Final Report. London: Her Majesty's Stationery Office, July 1963. pp. 173-85.

42. Moles, Abraham. Information Theory & Esthetic Perception. Joel E. Cohen (transl.) Urbana: U. of Illinois Press. 1966.

43. Parrack, H. "Community Reaction to Noise," Handbook of Noise Control. Cyril M. Harris (ed.) N.Y.: McGraw-Hill Book Co. 1957.

44. Proshansky, H. M., W.H. Ittelson, and L.G. Rivlin (eds.) Environmental Psychology: Man and his Physical Setting. N.Y.: Holt, Rinehart & Winston, 1970.

45. Tankin, Robert E. "Air Pollution Control & Public Apathy," J. of the Air Pollution Control Association. 19 Aug. 1969. 565-69.

46. Ruff, George E. "Adaptation under Extreme Environmental Conditions," The Annals of the American Academy of Political & Social Sciences, R. Lambert (ed.) May 1970. pp. 19-27.

47. Saarinen, Thomas F. Perception of Environment. Commission on College Geography, Resource Paper #5, Association of American Geographers, Washington, D.C. 1969.

48. Saroff, Jerome R., and Alberta Z. Lavitan. Survey Manual for Comprehensive Urban Planning: the Use of Opinion Surveys & Sampling Techniques in the Planning Process. Fairbanks, Alaska, Inst. of Social, Economic and Government Research, U. of Alaska. 1969. (SEG Report #19).

49. Sewell, W.R. Derrick. "Environmental Perceptions and Attitudes of Engineers & Public Health Officials," in Environment & Behavior, Vol. 3, No. 1. Mar. 1971. 23-59.

50. Shaffer, Margaret T. "Attitudes, Community Values and Highway Planning," in Transportation Impact & Attitude Surveys. Highway Research Record No. 187. 1967.

51. Shanley, R. A. Community Leader Attitudes on Water Pollution Abatement in Selected Massachusetts Communities on the Connecticut River. Publication #4 of the Water Resources Research Center, U. of Massachusetts, Amherst. 1968.

52. "Techniques for Determining Community Values " Record Number 102. Highway Research Board. 1965. pp. 11-18.

53. US EPA. Effects of Air Pollution on Public Attitudes and Knowledge. June, 1972. 151 pp.

54. Vernon, Magdalen D. The Psychology of Perception. Baltimore: Penguin Books, A530. 1962.

55. Zajonc, Robert B. "Attitudinal Effects of Mere Exposure." J. of Personality & Social Psychology. 9 June, 1968. pp. 1-27.

General Bibliographical Source

An Annotated Bibliography on Environmental Perception with Emphasis on Urban Areas. Council of Planning Librarians. 1969.

ECOLOGY

Ecology focuses on the interrelationships between living organisms and their environment. The recent heightened sensitivity of many peoples to the concept of environmental quality has assisted the development of ecology as an important branch of science recognized as relevant to everyday life. In order to comprehend the inherent complexities in this subject area, it is necessary to adopt a general and integrative approach. This approach is also necessary because ecology has evolved as an integrative science developing from a variety of disciplines including biology, geology, physiology, geography, meteorology, and many others. It is this interdisciplinary approach that has particular relevance to environmental management, as the study of ecology involves developing expertise for the purpose of evaluating the utilization of land, air and water habitats.

Since all human development activity takes place in some conjunction with natural systems, it is important to have a firm understanding of ecological limitations if our development is to be properly managed. Moreover, provision for the rational use of the earth's resource base implies consideration of the biological and physical precepts upon which all life functions. To fully appreciate the dynamic processes which may produce or act to disturb stability in the biosphere, it becomes essential to have a fundamental knowledge of underlying biological mechanisms.

Knowledge of these processes should assist a manager in asking penetrating questions about the ecological phenomena he confronts when performing environmental policy analysis. Some of these questions are: What basic ecological principles describe environmental dynamics? What are the relations between activities and effects on ecosystems? What kinds of responses can we expect from ecosystems when pollutants are introduced? How can planned developments be environmentally efficient? This subject area highlights consideration of these questions and many others.

ECOLOGY

General Outline

Basic Principles and Concepts of Ecology

- Structure
- Energy
- Biochemical Cycles
- Growth and Productivity
- Biological Transfers
- Threshold Factors
- Organization
- Adaptations to Perturbations and Stresses
- Time and Space
- Resource Utilization

Types of Ecosystems

- Terrestrial
- Freshwater
- Marine
- Estuarine

ECOLOGY

Detailed Outline

I. Basic Principles and Concepts of Ecology [32,40]

A. Principles and Concepts Pertaining to Ecosystem Structure [31,43]

1. Cell
2. Organism
3. Population
4. Community
5. Ecosystem

B. Principles and Concepts Pertaining to Energy [10,20]

1. Biotic energy flows
2. Food Chain
   a. autotrophic organisms
   b. heterotrophic organisms

C. Principles and Concepts Pertaining to Biogeochemical Cycles

1. Mineral cycles and nutrient regeneration
2. Nutrient exchange rate between organisms and the environment
3. Biogeochemical cycles
   a. the Carbon Cycle
   b. the Hydrologic Cycle
   c. the Sulphur Cycle [2]
   d. the Nitrogen Cycle [5,8]
   e. the Phosphorus Cycle

D. Principles of Growth and Productivity [12]

1. Growth cycles
2. Life cycles
3. Niche specialization

4. Size of organism

5. Metabolic rate

6. Physiology

7. Cell aging rate

8. Survival

E. Principles and Concepts Relating to Biological Transfers and Modifications [3,11,14,26]

1. Biological uptake

2. Biotransformation

3. Synergism

4. Response

F. Principles and Concepts Pertaining to Threshold Factors

1. Limiting factors affecting growth and productivity

a. abiotic factors (sample listing)

(1) light
(2) temperature
(3) salinity
(4) pH
(5) rate limiting nutrients[15]
(6) turbulent mixing
(7) advective transport

b. biotic factors (sample listing)

(1) predator-prey
(2) competition

2. Predictability of the occurrence of the limiting factor [28,30]

3. Tolerance limits [19]

G. Principles and Concepts Pertaining to Organization [23]

1. Population level

a. population size

(1) natality
(2) mortality
(3) dispersal

b. population dynamics [27]

(1) fluctuation
(2) competition
(3) parasitism

c. population control

2. Community Level

a. competition and predator-prey relationships

b. ecological succession

3. Ecosystem [36,37]

a. homeostasis

b. succession

H. Principles of Responses and Adaptations to Perturbations and Stresses in the Environment [6]

1. Perturbations introduced by weather

2. Perturbations and systems degradation caused by man

3. Stabilizing mechanisms [4,41]

4. Adaptations

a. physiological

b. behavioral

c. morphological

I. Principles Pertaining to Time and Space as Ecological Variables

1. Time as a resource limitation for all organisms

2. The structure of available space

J. Principles of Resource Use [1,25]

1. Resource availability [38]

2. Mechanisms providing for the availability of substances and the effect of matter on living organisms

3. Resource utilization

II. Types of Ecosystems [32,40]

- A. The terrestrial environment
  1. Forests [22,33,42]
     - a. tropical lowland Forests
     - b. National Forests
  2. Watersheds
  3. Range and grasslands [21]
     - a. savannas
     - b. temperate grasslands
     - c. semi-arid grasslands
  4. arid and semi-arid regions [7,34]
  5. agricultural
  6. high-latitude
- B. The freshwater environment [16]
  1. lentic community
  2. lotic community [18]
- C. The marine environment [9,29,35,39]
  1. major biomes
  2. sub-biomes
- D. The estuarine environment [13,24]
  1. geographic location
  2. type of organisms

ECOLOGY

Bibliography

Cited References

The following references are those cited by the accompanying outline:

1. Bakuzis, E.V. The Ecosystem Concept in Natural Resource Management. "Forestry Viewed in an Ecosystem Perspective" in Van Dyne, G.M. (ed.) Academic, New York. 1969. pp. 189-258.

2. Berner, R.A. Sulfate Reduction, Pyrite Formation, and the Oceanic Sulfur Budget. Paper presented at Nobel Symposium. Stockholm. 1971.

3. Bloom, Sandra C., and Stanley E. Degler. Pesticides and Pollution. Washington, D.C.: BNA Books. 1969. 99.

4. Brookhaven Symposia in Biology. Diversity and Stability in Ecological Systems. Brookhaven National Laboratory Assoc., Universities Inc., Upton, New York. 1969.

5. Commoner, B. "Threats to the Integrity of the Nitrogen Cycle: Nitrogen Compounds in Soil, Water, Atmosphere and Precipitation", in S.F. Singer (ed.), Global Effects of Environmental Pollution, D. Riedel, Dordrecht, Holland. 1970.

6. DeBach, Paul H., (ed.) Biological Control of Insect Pests and Weeds. New York: Van Nastrand Reinhold. 1964. 844.

7. Dregne, Harold E., (ed.) Arid Lands in Transition. Washington, D.C.: American Association for the Advancement of Science. 1970. 524.

8. Endelman et al. "A Systems Approach to an Analysis of the Terrestrial Nitrogen Cycle". J. Environ. Sys. Vol. 2(1). 3-19. 1972.

9. Friedrich, H. Marine Biology. University of Washington. 1969.

10. Gates, D.M. Energy Exchange in the Biosphere. New York: Harper and Row. 1962.

11. Gillett, J.W. (ed.) The Biological Impact of Pesticide in the Environment. Oregon State University, Corvalles, Oregon, 1970.

12. Goldman, C.R. Primary Productivity in Aquatic Environments. University of California. 1969.

13. Green, J. Biology of Estuarine Animals. University of Washington. 1958.

14. Harrison, et al. "Systems Studies in DDT Transport". Science. Vol. 170. pp. 503-508. 1970.

15. Hood, D. (ed.) Organic Matter in Natural Waters. University of Alaska, Inst. of Mar. Sci. 1970.

16. Hutchinson, C.E. A Treatise on Limnology. Vols. I and II, Wiley. 1967.

17. Hurd, L.E. "Stability and diversity at three tropic levels in terestrial successional ecosystems". Science. May 5, 1972.

18. Hynes, H.B.N. The Ecology of Running Waters. Univ. Toronto Press. 1970.

19. Lead Development Association. Lead Abstracts. Lead Development Association, 34 Berkeley Square London W1, England.

20. Lehninger, A.L. Bioenergetics. W.A. Benjamin Inc. 1971.

21. Lewis, J.K. "Range Management Viewed in the Ecosystem Framework" in G.M. Van Dyke (ed.) The Ecosystem Concept in Natural Resource Management. Academic, New York. 1969. pp. 97-187.

22. Little, Silas, and John H. Noyes. (eds.) Trees and Forests in an Urbanizing Environment. Amherst, Massachusetts: Cooperative Extension Service, University of Massachusetts Press. 1971. 168.

23. MacArthur, R.H. (ed.) Geographical Ecology: Patterns in the Distribution of Species, New York: Harper & Row, 1972.

24. McLusky, D. Ecology of Estuaries. New York: Hillary House Publishers Ltd.

25. National Academy of Sciences - National Academy of Engineers, National Research Council. Committee on Agriculture, Land Use, and Wildlife Resources. Land Use and Wildlife Resources. Washington, D.C.: National Academy of Sciences - National Academy of Engineers. 1970. 262.

26. Nelson-Smith, A. Oil Pollution and Marine Ecology. New York: Plenum. 1973. 256.

27. Nikolskii, G.V. Fish Population Dynamics. Oliver & Boyd. 1969.

28. Nilsson, R. "Aspects on the Toxicity of Cadmium and its Compounds" Swedish Natural Science Research Council, Bull. 7, Stockholm. 1970.

29. Nybakken, J.W. Readings in Marine Ecology. Harper and Row. 1971.

30. Oak Ridge Environmental Information System Office, Oak Ridge National Laboratory. Toxic Materials in the Environment.

31. Odum, E.P. "The Strategy of Ecosystem Development". Science, No. 164. 1969. pp. 262-270.

32. Odum, E.P. Fundamentals of Ecology. W.B. Saunders Co. 1971.

33. Reichle, David E. (ed.) Analysis of Temperate Forest Ecosystems. Ecological Studies 1. 1970.

34. Slayter, R.O. and R.A. Perry (eds.) Arid Lands of Australia. Australian National University Press, Canberra, A.C.T. 1969.

35. Steele, J.H. Marine Food Chains. Univ. of California. 1971.

36. Stumm, W. and E. Stumm. Chemostasis and Homeostasis in Aquatic Ecosystems: Principles of Water Pollution Control. Symp. on Non-Equil. Concepts of Natural Waters, American Chemical Society, Houston, Texas. Feb. 1970.

37. U.S. Congress Report. Diversity and Stability in Ecosystems. 1969.

38. Van Dyke, G.M. (ed.) The Ecosystem Concept in Natural Resource Management, Academia Press, New York. 1969.

39. Vernberg, W.B. and F.J. Vernberg. Environmental Physiology of Marine Animals. Springer-Verlag. 1972.

40. Watt, K. Principles of Environmental Science. McGraw-Hill Co. 1973.

41. Watt, K. E. A Comparative Study on the Meaning of Stability in Five Biological Systems: Insect and Furbearer Populations, Influenza, Thai. Hemorrhagic Fever, and Plague. Brookhaven Sym. in Biology, No. 12. 1969.

42. Weddle, Richard M. (ed.), Forest Land Use and the Environment. Missoula: School of Forestry, University of Montana. 1972. 150.

43. Wiens, John A. (ed.) Ecosystem Structure and Function. Oregon State Univ. Press. 1972. 176 pp.

44. Wilber, C.C. Biological Aspects of Water Pollution. Charles C. Thomas Publishers. 1969.

General References

The following references are of general interest for this subject area:

Alexander, M. Microbial Ecology. Wiley & Sons. 1971.

American Chemical Society, Committee on Chemistry and Public Affairs. Cleaning Our Environment, the Chemical Basis for Action. American Chemical Society, Washington, D.C. 1969.

Barnes, R.D. Invertebrate Zoology. 2nd Ed., Saunders. 1968.

Cole, Lamont C. "Man's Ecosystem". Environments of Man. Addison-Wesley, Reading, Mass. 1968. p.3-16.

Colinvaux, Paul A. Introduction to Ecology. New York: John Wiley. 1972.

Dasmann, R. F. et al. Ecological Principles for Economic Development. John Wiley Ltd. 1973.

Dasmann, R.F. Environmental Conservation. 3rd ed., R.F. Dasmann, New York: John Wiley. 1972.

Detwyler, T.R. Man's Impact on Environment. New York, McGraw-Hill. 1971. 731 pp.

Directory of USAEC Information Analysis Centers. Ecology and Environmental Quality Bibliography. Syracuse, N.Y., Syracuse University Library. Jan. 1972.

Dugan, P. Biochemical Ecology of Water Pollution. New York: Plenum Publishing Corp.

Edwards, C.A. (ed.) Environmental Pollution by Pesticides. New York: Plenum. 1973. 480.

Ehrlich, Paul et al. Human Ecology. W.H. Freeman & Co., San Francisco. 1973.

Englewood, C. Environmental Crisis: Will We Survive? Englewood Cliffs, N.J.: Prentice-Hall. 1972.

Erodus and Morgan. A Nationwide Survey of Environmental Protection. Erodus and Morgan, Inc., New York: The Wall Street Journal. 1972.

Faidman, C. and J. White. (eds.) Ecocide ... and Thoughts Toward Survival. New York: Interbook. 1972.

Fogg, C.E. Algal Cultures and Phytoplankton Ecology. University of Wisconsin Press. 1965.

Hall, G. Ecology: Can We Survive Under Capitalism? New York: International Publishers. 1972.

The Institute of Ecology. Man in the Living Environment: Report of the Workshop on Global Ecological Problems. Madison, Wisconson. 1972.

IUCN Bulletin. "Mammals Listed in the Red Data Book: Vol. 1". IUCN Bulletin, April 1973, V4, n4(8) survey report.

Johnson, Huey D., (ed.) NO DEPOSIT-NO RETURNS: Man and His Environment: A View Toward Survival. Addison-Wesley Publishing Company. 1970.

Johnson, G.L., and C.L. Quance (eds.) Overproduction Trap in U.S. Agriculture: A Study of Resource Allocation Between World War One to the Late 1960's. Baltimore: Johns Hopkins Press. 1972.

Krebs, C.J. Ecology: The Experimental Analysis of Distribution and and Abundance. Harper & Row. 1972.

Kormondy, Edward J. Concepts of Ecology. Prentice-Hall, Inc., Englewood Cliffs, N.J. 1969.

Matthews, William H. et al. (ed.) Man's Impact on Terrestrial and Oceanic Ecosystems, Cambridge, Mass: MIT Press. 1972.

McHarg, Ian. Design with Nature. Doubleday & Co., Inc. 1971.

Michigan State University Press. Pesticides in the Soil: Ecology Degradation, and Movement. Michigan State University Press, East Lansing, Mich. 1970.

Mitchell, R. Water Pollution Microbiology. Wiley-Interscience. 1972.

Myrup, L.O. "A Numerical Model of the Urban Heat Island". J. Appl. Meteorol. Vol. 8, 1969. pp. 908-918.

Noble, Philip and John Deedy (eds.) The Complete Ecology Fact Book. Garden City, N.Y.: Doubleday. 1972.

Odum, Howard Thomas. Environment, Power, and Society. 1971.

Olson, T.A. and Burgess, E.J. Pollution in Marine Ecology.

Paddock, W.C. "How Green is the Green Revolution?" Bioscience, No. 20, 1970. pp.897-902.

Patten, B.C. Systems Analysis and Simulation in Ecology. Vol. 1, Academic Press. 1971.

Patterson, C.C. and J.D. Salvia. "Lead in Natural Environment". Science, no. 159. 1968. p. 1000.

Peterson, A. PCB - Possibilities of Destruction. PCB Conference Wenner-Gren Centre, Stockholm. 1970.

Rehoe, R.A. "Lead Intake from Food and Water". Science, no. 159. 1968. p. 1000.

Ricklefs, R.E. Ecology. Chiron Press. 1973.

Risebrough, R.W., P. Rieche, D.B. Peakall, S.G. Herman and M.N. Kirven. "Polychlorinated Biphenyls in the Global Ecosystem". Nature, no. 220, pp. 1098-1102.

Sangster, R. P. Ecology: A Selected Bibliography. #170, Council of Planning Librarians. 1971.

Schultz, A. M. Ecosystem and Environment. Canfield Press. 1972.

Strickland, J. D. H. Microbial Activity in Aquatic Environments. Institute of Marine Resources, La Jolla, California.

Study of Critical Environmental Problems (SCEP). Man's Impact on the Global Environment: Assessment and Recommendations for Action. Cambridge, Mass.: MIT Press. 1970.

Stumm, W. and J. J. Morgan. Aquatic Chemistry. Wiley-Interscience, 1970.

Swedish Natural Science Research Council. Ecosystem Approach to the Baltic Problem. Written by Bent-Owe Jansson, Bulletin #16 from NFR, Stockholm. 1972.

Thomann, R. V. Systems Analysis and Water Quality Management. Environmental Research and Application. New York. 1971.

Turk, A. et al. Ecology, Pollution, Environment. Philadelphia: W. B. Saunders. 1972.

United Nations. Food and Agriculture Organization. Fish and Fisheries in the Context of Environmental Concerns. Paper prepared for the United Nations Conference on the Human Environment, Stockholm, Sweden, June 5-16, 1972. New York: United Nations, Food and Agriculture Organization. 1971. 13.

United States President's Commission on Marine Science, Engineering and Resources. Our Nation and the Sea. Washington, D.C.: United States Government Printing Office. 1969. 305.

United States Department of the Interior. Bureau of Reclamation. Ecological Impact of Water Resource Development. Washington, D.C.: United States Government Printing Office. 1972. 28.

United States Congress. Senate. Committee on Interior and Insular Affairs. Subcommittee on Public Lands. "Clear-Cutting" Practices on Timberlands. Hearings of the Subcommittee before the 92nd Congress, 1st Session, April 5-June 29, 1971. Washington, D.C.: United States Printing Office. 1971. (3 parts). 1247.

United States Congress. Senate. Committee on Commerce. Subcommittee on Energy, Natural Resources and the Environment. Effects of 2,4,5-T on Man and the Environment. Hearings of the Committee before the 91st Congress, 2nd Session, April 7, 15, 1970. Washington, D.C.: United States Government Printing Office. 1968. 13. (House Report 1223).

United States Congress. House of Representatives. Committee on Merchant Marine and Fisheries. Environmental Data Bank. Hearings before the Subcommittee on Fisheries and Wildlife Conservation of the Committee on Merchant Marine and Fisheries. 1970.

Van Dyke, G.M. Ecosystems, Systems Ecology, and Systems Ecologists. U.S. Atomic Energy Comm., Oak Ridge Nat'l Lab, Report #3957. 1966.

Ward, B. and Dubos, R. Only One Earth: The Core and Maintenance of a Small Planet. New York: W.W. Norton & Company, Inc.

Warren, C.E. Biology and Water Pollution Control. Saunders. 1971.

Winton, H. Man and the Environment, A Bibliography of Selected Publication of the United Nations System. 1946-1971. R.R. Bowker Co., 1972.

Woodwell, G.M. "Effects of Pollution on the Structure and Physiology of Ecosystems." Science, 168: 3930: 429-433. April 24, 1970.

Wurster, C.F. "Chlorinated Hydrocarbon Insecticides and the World Exosystem" Biol. Conserv. 1969 pp.123-129.

ENVIRONMENTAL EFFECTS

This subject area relates environmental effects to their causes. Associated with an activity is a set of fairly obvious direct or primary effects. However, the influence of the activity does not often terminate at these effects; instead, there is often a succession of derivative or secondary or indirect effects that follow one another like the links of a chain. Complicating the picture of the cause-and-effect relationship is the fact that an effect is generally not the result of one cause alone but of several converging causes. Also any effect may generate a number of direct and indirect effects which will occur at different times and in different places. An environmental manager should not infer from the structure of this subject area that cause and effect is a one-to-one relationship. A network is a better representation of the interaction and relationship among effects.

The purpose of this subject area is to expose an environmental manager to the complexity of the environment. However, even by broadening an environmental manager's perspectives on the spectrum of effects an activity may have on the environment, this subject area alone will not enable an environmental manager to contribute effectively to the preservation and enhancement of the environment, and at the same time achieve other objectives. After identifying all the probable effects of an activity, an environmental manager must be acquainted with what environmental indicators he can use to measure these effects, be aware of what values he should be concerned with, and be able to assess the total costs and benefits of the activity. Treatment of these subjects is provided in the subject areas of Environmental Indicators, Values and Perception, and

Environmental Impact Assessment Methodology respectively. It must be emphasized that all subject areas are in some way related to one another, and the greatest use can be made out of each subject area only if the manager studies them all.

ENVIRONMENTAL EFFECTS

General Outline

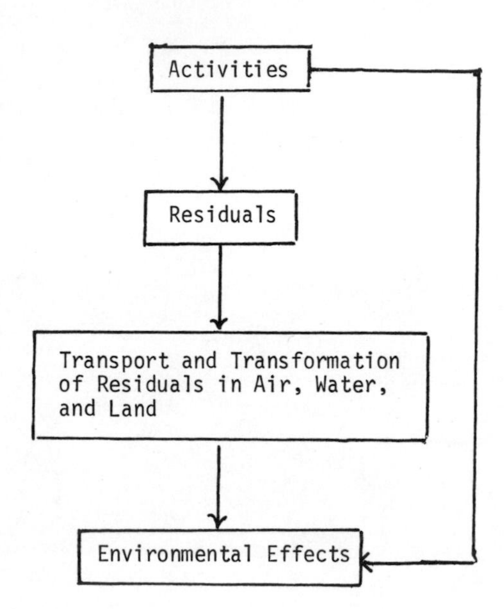

ENVIRONMENTAL EFFECTS

Organization of Outline

I. Activities vs. Residuals vs. Effects

A. Matrix relationships

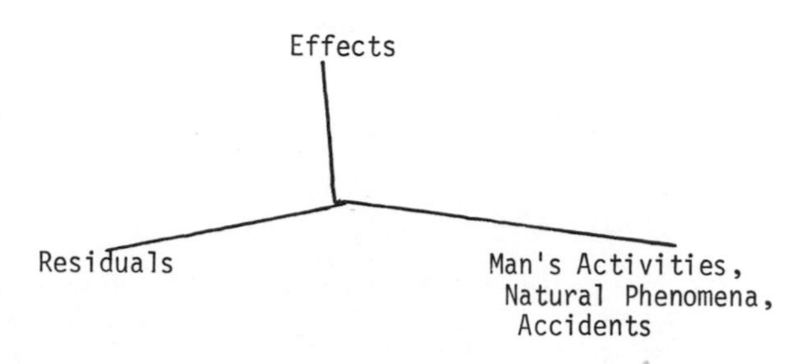

B. Residuals vs. Activities

Residuals	Section of Outline
Radiation	V.-A,B,
Noise	IV.-A
Solid Wastes	III.-A,B,C
Air Pollutants	II.-A,B,C
Water Pollutants	I.-A,B,C

C. Residuals vs. Effects

Residuals	Section of Outline
Radiation	V.-B,C
Noise	II.-B,C,D,E
Solid Wastes	III.-B,C,E
Air Pollutants	II.-C,E,F,G,H; VI.-B,D
Water Pollutants	I.-B,C,E,F,G,H

D. Effects vs. Activities

Effects on	Section of Outline
Man	VIII
Animals	
Plants	
Inanimate Objects	
Weather and Climate	VI.-A,C,D
Resources	VII.-A,B
Ecosystem	IX

II. Transport and Fate of Residuals

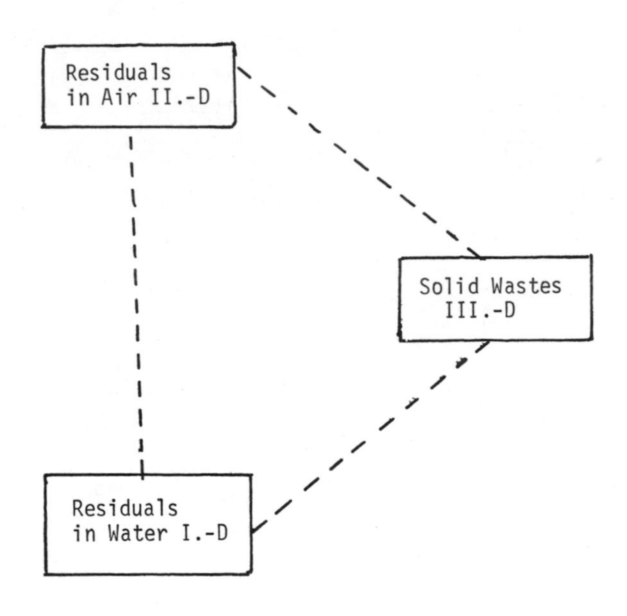

--- Intermedia Transport

III. Man's Activities, Accidents, and Natural Phenomena

A. Man's Activities

1. Agricultural, Forestry, and Fisheries
2. Chemical Treatment
3. Conservation and Recycling
4. Construction
5. Domestic
6. Manufacturing
7. Military
8. Power Generation
9. Recreation
10. Services
11. Scientific and Technological Research
12. Transportation
13. Waste Collection, Treatment, and Disposal
14. Weather Modification

B. Accidents

1. Explosion
2. Fire
3. Operational Failure
4. Spills and Leaks

C. Natural Phenomena

1. Earthquake
2. Flooding
3. Tornado
4. Tropical Storm
5. Tsunami
6. Volcanic Eruption

ENVIRONMENTAL EFFECTS

Detailed Outline*

I. Water Pollution [6,16,26]

A. Sources of water pollutants [36,51]

1. Direct discharge of effluent from:

a. Industries

b. Power generation [45]

c. Domestic sewage [40]

d. Surface runoff [40]

e. Ocean Dumping

f. Agricultural activities [1,12,33,42,50,55]

(1) application of pesticides and other chemicals
(2) irrigation return

g. Chemical treatment

h. Spills and leaks [7,8,11,15,17,21,30,31,37,41,46]

i. Wastes disposal [23,24,25,39,49]

2. Changes of water quality resulting from artificial manipulation of the flow pattern of water bodies

a. Dam construction

b. Channelization

c. Dredging [56]

d. Land transformation and other construction activities

e. Flow reversion

f. Ground water recharge

---

* Because of the large number of references for this subject area, bibliographies are provided at the end of each major section. The numbers in each section refer only to the bibliography at the end of that section. Some general references are given at the end of the subject area.

B. Water pollutants

1. Pollutants in groundwater, lakes, rivers, and estuaries

a. Infectious agents - microbial or viral

b. Oxygen-demanding wastes

c. Plant nutrients

(1) nitrogenous
(2) phosphorus
(3) carbonic

d. Organic chemicals [29,52]

(1) oil film
(2) color pigment
(3) other organic compounds

e. Inorganic chemicals [27,53]

(1) acidity constituents
(2) alkalinity constituents
(3) hardness constituents
(4) color pigment
(5) taste constituents
(6) odorous compounds

f. Heavy metals[9,10,22,32,38]

g. Sediments and suspended solids

h. Radioactive wastes [35]

i. Heat - stratification [5,13,45]

j. Economic poisons [48,50]

2. Pollutants in ocean

a. Dredge soil

b. Industrial wastes

c. Sewage sludge

d. Construction and demolition debris

e. Explosives

f. Chemical munitions

g. Radioactive wastes

h. Spill oils

(1) ocean shipping
(2) offshore drillings
(3) accidents

i. Others

C. Water bodies into which pollutants are discharged

1. Groundwater

2. Rivers

3. Lakes

4. Estuaries and marines [28,34,43]

5. Ocean [44]

D. Transport and the fate of water pollutants

1. Factors affecting the transport and transformation of water pollutants

a. Characteristics of water pollutants

(1) physical state
(2) settleability
(3) chemical reactivity

b. Source effects

(1) Nature of source
(a) intermittent
(b) continuous
(c) mobile
(d) stationary
(2) configuration of source
(a) point
(b) line
(c) area
(d) volume
(3) effluent temperature
(4) effluent exit velocity
(5) concentration and volume of effluent

c. Hydrologic factors

(1) vertical temperature profile
(2) velocity and volume of flow
(3) degree of mixing
(4) tidal action
(5) water table variation

2. Transformation and the fate of water pollutants

   a. Physical reaction

      (1) sedimentation
      (2) coagulation
      (3) adsorption

   b. Chemical reaction - oxidation, reduction, synthesis, and decomposition

   c. Biological uptake

E. Effects on man

1. Human health

   a. Means through which water pollutants adversely affect human health

      (1) drinking water contamination [14,18]
      (2) intake of food which accumulate water pollutants
      (3) direct contact with contaminated water - recreation

   b. Factors affecting the effect of water pollution on human health

      (1) medical status of the individual
      (2) effect of age and sex
      (3) synergism and antagonergism
      (4) amount of intake

   c. Adverse effects on human health

      (1) toxic
      (2) pathologic, chronic
      (3) irritating sensory organs

2. Nuisance and general opposition -- palatability of drinking water [14,18,47]

   a. Odor

   b. Taste

   c. Turbidity

   d. Freshness and coolness

3. Aesthetics
    a. Oil film
    b. Color
    c. Foam
    d. Turbidity
4. Recreation -- undesirable for recreational usage because of adverse health effects and aesthetic concerns

F. Effects on wildlife, domesticated animals, and fishery products [4,5,54]
1. Behavioral patterns
2. Disease transmission
3. Malformation
4. Population alteration
    a. Migration
    b. Mortality
    c. Reproduction
5. Disorders

G. Effects on vegetation and crops [4,54]
1. Vegetation damage and reduced crop yield
    a. Toxic
    b. Chronic
2. Eutrophication [3,19,20]

H. Effects on inanimate objects
1. Objects vulnerable to damage
    a. Water network
    b. Boiler and heating equipment
    c. Paper and allied products
    d. Textile and textile products

2. Adverse effects

   a. Scaling

   b. Coloring

   c. Corrosion

   d. Imparting impurities into products

Water Pollution

Bibliography

1. Agricultural Pollution of the Great Lakes Basin. Water Quality Office, Environmental Protection Agency, Chicago, Ill. 1 July, 1971. 186 pp.

2. Atlas, Ronald M. Fate and Effects of Oil Pollutants in Extremely Cold Marine Environments. Jet Propulsion Lab., Pasadena, Calif. 1 Nov. 1973. 37 pp.

3. Bartsch, A.F. Role of Phosphorus in Eutrophication. National Environmental Research Center, Office of Research and Monitoring, U.S. Environmental Protection Agency. Program Element 328201, EPA-R3-72-001.

4. Battele's Columbus Laboratories. Water Quality Criteria Data Book, Vol. 3: Effects of Chemicals on Aquatic Life Selected Data from the Literature through 1968. U.S. Environmental Protection Agency. May 1971.

5. Becker, C. Dale. "Columbia River Thermal Effects Study: Reactor Effluent Problems." Journal of the Water Pollution Control Federation. May 1973. V.45. n5. p.850 (20).

6. Benefits of Water Quality Enhancement. Dept. of Civil Engineering, Syracuse Univ., N.Y. Dec. 1970. 194 pp.

7. Blumer, Max. "Oil Pollution of the Ocean." From Man's Impact on Environment. McGraw-Hill Book Co. 1971. pp. 295-301.

8. Boesch, Donald F., Carl H. Hershner and Jerome Milgram. Oil Spills and the Marine Environment. Ballinger Publishing Co., Cambridge, Mass. June 1974. 112 pp.

9. Button, D.K., and S.S. Dunker. Biological Effects of Copper and Arsenic Pollution. Inst. of Marine Science, Alaska Univ., College, Alaska. Apr. 1971. 61 pp.

10. California Dept. of Public Health, Interagency Committee on Environmental Mercury. Mercury in the California Environment. California Dept. of Public Health, Environmental Health and Consumer Protection Program. 1971.

11. Chan, Gordon. The Effects of the San Francisco Oil Spill on Marine Life. National Technical Information Service, U.S. Dept. of Commerce. Jan. 1972. 78 pp.

12. Characteristics and Pollution Problems of Irrigation Return Flow. Utah State Univ., Logan, Utah. May 1969. 250 pp.

13. Christianson, Alden G., and Bruce A. Tichenor. Industrial Waste Guide on Thermal Pollution. Pacific Northwest Water Lab., Federal Water Pollution Control Administration, Corvallis, Oreg. Sept. 1968. 121 pp.

14. Collins, Ralph P. Characterization of Waste and Odors in Water Supplies. Biological Sciences Group, Connecticut Univ., Storrs, Connecticut. Aug. 1971. 20 pp.

15. Cowell, E.B. The Ecological Effects of Oil Pollution on Littoral Communities: Proceedings of a Symposium Organized by the Institute of Petroleum and held at the Zoological Society of London, November 30 - December 1, 1970. Applied Science Publihsers, Essex, England. 1971. 250 pp.

16. Crosslink, et al. The Value of the Tidal Marsh. Urban and Regional Development Center, Univ. of Florida. May 1973.

17. Degler, Stanley E. (ed.) Oil Pollution: Problems and Policies. Washington, D.C.: Bureau of National Affairs, Inc. 1969. 142 pp.

18. European Standards for Drinking Water. World Health Organization, Geneva. 1960.

19. Eutrophication of Surface Waters -- Lake Tahoe. Lake Tahoe Area Council, S. Lake Tahoe, Calif. May 1971. 157 pp.

20. Eutrophication of Surface Waters -- Lake Tahoe Indian Creek Reservoir. Lake Tahoe Area Council, S. Lake Tahoe, Calif. July 1971. 117 pp.

21. Foster, A., M. Neushul, A. C. Charters, and R. Zingmark. Santa Barbara Oil Pollution, 1969. California Univ., Santa Barbara, California. Oct. 1970. 56 pp.

22. Friberg, Lars, Gosta Lindstedt, Gunnar Nordberg, Claes Ramel, and Staffan Skerfving. Mercury in the Environment. A Toxicological and Epidemiological Appraisal. Dept. of Environmental Hygiene, Karolinska Institutet, Stockholm, Sweden. Nov. 1971. 551 pp.

23. Fungaroli, A.A. Pollution of Subsurface Water by Sanitary Landfills. Vol. I. Drexel Univ., Philadelphia, Pa. 1971. 198 pp.

24. Fungaroli, A.A. Pollution of Subsurface Water by Sanitary Landfills. Vol. II. Drexel Univ., Philadelphia, Pa. 1971. 221 pp.

25. Fungaroli, A.A. Pollution of Subsurface Water by Sanitary Landfills. Vol. III. Drexel Univ., Philadelphia, Pa. 1971. 174 pp.

26. Furon, Raymond, Translated by Paul Barnes. The Problems of Water: a World Study. Faber and Faber Ltd., London. 1963. 175 pp.

27. Gammon, James R. The Effect of Inorganic Sediment on Stream Biota. Depauw Univ., Greencastle, Ind. Dec. 1970. 150 pp.

28. Goldberg, E.G. A Guide to Marine Pollution. NTIS reprint AD-751 198. Oct. 1972. 24 pp.

29. Goldberg, Edward D., Philip Butler, Paul Meier, David Menzol, David Paulik, Robert Risebrough, Lucille F. Stickel. Chlorinated Hydrocarbons in the Marine Environment. National Academy of Sciences, Committee on Oceanography, Panel on Monitoring pesticides in the marine environment. 1971. 47 pp.

30. Hepple, Peter (ed.) Water Pollution by Oil. Proceedings of Seminar held at Aviemore, Invernesshire, Scotland, sponsored by the Institute of Water Pollution Control and the Instituteof Petroleum with the assistance of the European Office of the World Health Organization. 4-8 May 1970. 393 pp.

31. Hoult, David P. (ed.) Oil on the Sea: Proceedings of a Symposium on the Scientific and Engineering Aspects of Oil Pollution of the Sea. Sponsors: Massachusetts Institute of Technology, Woods Hole Oceanographic Institute. Plenum Press. 1972. 114 pp.

32. Lepple, Frederick. Mercury in the Environment. Sea Grant Program, College of Marine Studies, U. of Delaware. Mar. 1973. 75 pp.

33. Loehr, Raymond C. Pollution Implications of Animal Wastes. A Forward Oriented Review. Dept. of Civil Engineering, Kansas Univ. Lawrence, Kansas. July 1968. 188 pp.

34. National Academy of Sciences, National Council on Ocean Affairs Board. Marine Environmental Quality: Suggested Research Programs for Understanding Man's Effect on the Oceans. National Academy of Sciences. 1971. 115 pp.

35. National Academy of Sciences. Radioactivity in the Marine Environment. Prepared by the Panel on Radioactivity in the Marine Environment of the Committee on Oceanography, National Research Council, National Academy of Sciences, Washington, D.C. 1971. 272 pp.

36. Nonpoint Rural Sources of Water Pollution. NTIS Report PB-214 508/4. 1972. 40 pp.

37. Ross, William M. Oil Pollution as an International Problem: A Study of Puget Sound and the Straight of Georgia. 1973. 224 pp.

38. Singer, Philip C. Trace Metals and Metal-Organic Interactions in Natural Waters. Ann Arbor Science Publishers, Inc. 1973. 380 pp.

39. Smith, David D., Robert P. Brown. Ocean Disposal of Barge-delivered Liquid and Solid Wastes from U.S. Coastal Cities. Applied Oceanography Division Dillingham Corp., U.S. Environmental Protection Agency. 1971. 119 pp.

40. Storm Water Pollution from Urban Land Activity. Avco Economic Systems Corp., Washington, D.C. 1969. 342 pp.

41. The Oil Spill Problem. First Report of the President's Panel on Oil Spills, Executive Office of the President, Office of Science and Technology, Washington, D.C. July 1970. 32 pp.

42. The Relationship Between Animal Wastes and Water Quality. A Report of Recent Meetings, October 1971 and January 1972. President's Water Pollution Control Advisory Board, Washington, D.C. 1972. 33 pp.

43. United Nations, Food and Agriculture Organization. Marine Pollution and its Effects on Living Resources. FAO Fisheries Report #99. June 1971.

44. U.N. Stockholm Conference on Ocean Pollution. Pollutants in the Ocean. Background paper.

45. United States Atomic Energy Commission. Thermal Effects of Projected Power Growth. The National Outlook.

46. U.S. Council on Environmental Quality. National Oil and Hazardous Substances Pollution Contingency Plan. U.S. Government Printing Office, Wash., D.C. Aug. 1971. 70 pp.

47. U.S. Dept. of Health, Education, and Welfare. Drinking Water Standards 1962 (Reprinted 1969). U.S. Dept. of Health, Education, and Welfare, Public Health Service, Consumer Protection and Environmental Health Service, Env. Control Administration, Rockville, Md.

48. U.S. Dept. of Health, Education, and Welfare. Report of the Secretary's Commission on Pesticides and their Relationship to Environmental Health. Parts I & II. U.S. Dept. of Health, Education, and Welfare, Wash., D.C., 1971. 677 pp.

49. U.S. Environmental Protection Agency. Pollution of Subsurface Water by Sanitary Landfills. Vol. I. U.S. Government Printing Office, Wash., D.C. 1971. 132 pp. + appendices.

50. U.S. Environmental Protection Agency, Office of Water Programs. The Effects of Agricultural Pesticides in the Aquatic Environment, Irrigation Croplands, San Joaquin Valley. U.S. Environmental Protection Agency. June 1972.

51. U.S. House of Representatives. Non-Point Source Pollution from Agricultural, Rural, and Developing Areas. Hearing House Committee on Public Works, 92 Congress II, Serial 92-49, Aug. 15-17, 1972.

52. Water Quality Criteria Data Book. Vol. I. Organic Chemical Pollution of Freshwater. Little (Arthur D.), Inc., Cambridge, Mass. Dec. 1970. 399 pp.

53. Water Quality Criteria Data Book. Vol. II. Inorganic Chemical Pollution of Freshwater. Little (Arthur D.), Inc., Cambridge, Mass. July 1971. 280 pp.

54. Wilber, Charles G. The Biological Aspects of Water Pollution. Springfield, Illinois: Charles C. Thomas. 1971. 296 pp.

55. Willrich, Ted L. and George E. Smith. Agricultural Practices and Water Quality. The Iowa State U. Press, Ames, Iowa. 1970.

56. Windom, Herbert L. Environmental Response of Salt Marshes to Deposition of Dredged Material. Skidaway Institute of Oceanography (presented at ASCE National Water Resources Engineering Meeting, Atlanta, Jan. 24-28, 1972). Jan. 1972.

General Bibliographical Sources on Water Pollution

Brinn, D.G. A Selected Bibliography on Pollution of Estuarine and Coastal Waters with Particular Regard to Industrial Effluents. British Iron & Steel Association. 1972.

Ditsworth, George R. Environmental Factors in Coastal and Estuarine Waters Bibliographic Series -- Vol. I. Coast of Oregon. Pacific Northwest Water Lab., Corvallis, Oregon. Oct. 1966. 70 pp.

Ditsworth, George R. Environmental Factors in Coastal and Estuarine Waters Bibliographic Series -- Vol. II. Coast of Washington. Pacific Northwest Water Lab., Corvallis, Oreg. Aug. 1968. 88 pp.

Florida. Department of Natural Resources, Coastal Coordinating Council. A Selected Bibliography on: Thermal Pollution; Thermal Effluents; and Electric Power Plants, their Effects, Planning and Siting. 1972 Supplement. Oct. 1972. 13 pp.

Office of Water Resources Research. Estuarine Pollution: A Bibliography. Office of Water Resource Research, Water Resources Scientific Information Center. PB-220 119/2. Apr. 1973. 510 pp.

Radcliffe, Donna R. and Thomas A. Murphy. Biological Effects of Oil Pollution -- Bibliography. A Collection of References Concerning the Effects of Oil on Biological Systems. Federal Water Pollution Control Administration, Wash., D.C. Oct. 1969. 52 pp.

Shih, H.H. A Literature Survey of Ocean Pollution. Catholic U. of America, NTIS report AD-743 101. May 1971. 171 pp.

Thermal Effects Literature. ORNL-NSIC-110 230 pp.

U.S. Office of Water Resources Research. Detergents in Water: A Bibliography. WRSIC 71-214. Dec. 1971.

U.S. Office of Water Resources Research. Mercury in Water: A Bibliography. WRSIC 72-207. Jan. 1972.

U.S. Office of Water Resources Research. PCB in Water: A Bibliography. WRSIC 72-201 Jan. 1972.

II. <u>Air Pollution</u> [5,13,24,29,30,45,46,52,66,67,86,100,110]

A. Sources of Air Pollution [3,4,85,124,132]

1. Man's activities

a. Agricultural activities

(1) crop spraying and dusting
(2) milling

b. Construction

c. Domestic

(1) domestic space heating
(2) commercial space heating

d. Manufacturing [6,88,96]

(1) chemical process industry [20,21,23,27,35,39,53,62]
(2) metallurgical industry [38,98,121,122,123]
(3) mineral products industry
(4) petroleum industry [22]
(5) textile industry
(6) wood processing [60]

e. Power generation

(1) fossil fuel combustion [40,65,101,102]
(2) nuclear reactor

f. Scientific and technological research - nuclear device testing

g. Transportation

(1) automobile [32,111,112,131]
(2) locomotive
(3) vessels
(4) aircraft [15,73,89,90]

h. Waste disposal [61]

(1) incineration
(2) open burning

2. Accidents

a. Forest fire

b. Operation failure

3. Natural phenomena

   a. Volcanic eruption

   b. Meteors

   c. Sea spray

   d. Bacterial action

   e. Natural chemical reaction

B. Processes generating air pollutants

1. Combustion

2. Spraying

3. Evaporation loss

C. Air Pollutants

1. Inorganic gases

   a. Carbon compounds

      (1) carbon monoxide [7]
      (2) carbon dioxide

   b. Nitrogenous compounds [9]

      (1) nitric oxide
      (2) nitrogen dioxide
      (3) nitrogen pentoxide
      (4) ammonia [69]

   c. Ozone and photochemical oxidants [11]

   d. Sulfur compounds [12]

      (1) sulfur dioxide [72]
      (2) sulfur trioxide [72]
      (3) carbon disulfide
      (4) hydrogen sulfide [71]

2. Organic gases and vapors

   a. Hydrocarbons [8]

      (1) paraffins [105]
      (2) olefins
      (3) aromatics

b. Oxygenated hydrocarbons

(1) aldehydes [103]
(2) ketone
(3) alcohol
(4) organic acids
(5) phenols
(6) organic hydroperoxides and peroxides

c. Organic sulfur compounds

d. Organic nitrogenous compounds

(1) basic nitrogenous compounds
(2) peroxyl acetyl nitrate (PAN)
(3) amines

e. Chlorinated organic compounds

f. Organo-metallic compounds

g. Economic poisons

3. Particulates [10,63,99,128]

a. Mist

(1) sulfur trioxide and sulfuric acid
(2) hydrogen fluoride
(3) hydrochloric acid [104,106,126]
(4) alkali fumes
(5) oil and gasoline mists

b. Solids

(1) fly ash
(2) carbon
(3) asbestos [25,76,120]
(4) arsenic
(5) beryllium [25,41]
(6) cadmium [16,51,74]
(7) colloidal silica
(8) fluorides [28]
(9) iron oxides [116]
(10) lead
(11) mercury [25,107]
(12) phosphorus [17]
(13) vanadium [18]
(14) zinc [19,81]
(15) boron [42,78]

(16) barium [70,77]
(17) nickel [75,118]
(18) copper [79]
(19) selenium [80,108]
(20) chromium [115]
(21) manganese [117]

4. Others

a. Pollens and spores [48,49]

b. Pesticide drift [50]

c. Microbial agents

d. Seasalt nuclei

e. Combustion nuclei

f. Ground talc

g. Allergens

h. Radioactive substances

i. Other odorous compounds [83,84,119]

D. Transport and the fate of air pollutants

1. Factors affecting the transport and transformation of air pollutants

a. Characteristics of air pollutants

(1) physical state
(2) particle size
(3) chemical reactivity

b. Source effects

(1) nature of source
(a) intermittant
(b) continuous
(c) mobile
(d) stationary
(2) configuration of source
(a) point
(b) line
(c) area
(d) volume
(3) effluent temperature
(4) effluent exit velocity
(5) concentration and volume of effluents

c. Meteorological factors

(1) wind and turbulence
(2) humidity
(3) solar radiation
(4) vertical temperature profile

d. Topography of airshed

2. Transformation and the fate of air pollutants

a. Physical reaction

(1) coagulation
(2) adsorption
(3) solution

b. Chemical reaction

(1) stabilization through oxidation, reduction, synthesis, and decomposition
(2) photochemical reaction

c. Sink mechanisms

(1) remains in the atmosphere in the stabilized form
(2) gravitational fall-fallout
(3) meteorological precipitation - washout
  (a) by rain and drizzle
  (b) by snow and sleet

E. Effects of air pollution on human health [26,37,58,59,64,94,95,109]

1. Air pollutants having adverse effects on human health

a. Arsenic [114]

b. Asbestos

c. Beryllium

d. Cadmium

e. Carbon monoxide

f. Formaldehyde

g. Hydrogen fluoride [56]

h. Hydrocarbons

i. Hydrogen sulfide

j. Inorganic particulates

k. Lead [44,125]

l. Mercaptan

m. Mercury

n. Nitrogen dioxide [33,54]

o. Organic particulates

p. Ozone

q. Peroxyacyl nitrate (PAN)

r. Sulfur dioxide [127]

s. Sulfur trioxide [2,127]

t. Total oxidant

u. Total sulfate

v. Suspended particulate and settleable matter

w. Hydrogen chloride [55]

2. Target organs vulnerable to air pollutant attack

   a. Skin

   b. Mucous membranes

   c. Conjunctivae of the eye

   d. Linings of the respiratory tract and lungs [133]

   e. Linings of the digestive tract

3. Factors affecting the effect of air pollutants on human health

   a. Concentration - time relations

   b. Immunity developed by the individual from previous exposure

   c. Medical status of the individual

   d. Effect of age and sex

   e. Synergism and antagonergism

4. Effects on human health

   a. Acute sickness or death - episodes [1]

   b. Insidious or chronic disease [2]

      (1) attack on respiratory system - bronchitis, emphysema [43,133]
      (2) shortening of life
      (3) impairment of growth
      (4) genetic effect - carcinogenic, mutagenic, teratogenic [31,87]

   c. Alteration of important physiological functions

      (1) ventilation of lungs - asthma
      (2) transport of blood by haemoglobin
      (3) dark adaptation
      (4) other functions of nervous system

   d. Untoward symptoms

      (1) sensory irritation - eye, nose, throat, cough, headache, sinus
      (2) allergy

   e. Discomfort - distress, odor, impairment of visibility

F. Effects of air pollutants on animals [34,93]

1. Air pollutants having adverse effects on animals

   a. Arsenic

   b. Beryllium

   c. Cadmium

   d. Formaldehyde

   e. Hydrogen fluoride

   f. Mercury

   g. Nitrogen oxides

   h. Peroxyacyl nitrate (PAN)

   i. Sulfur dioxide

   j. Total oxidant

2. Adverse effects on animals

   a. Toxic

   b. Chronic

   c. Irritating

G. Effects of air pollutants on vegetation [57,68,91,97]

1. Pollutants found in the atmosphere with concentration sufficient to cause adverse effects on vegetation

   a. Sulfur dioxide

   b. Sulfuric acid mist

   c. Hydrogen sulfide

   d. Sulfate

   e. Nitrogen oxides

   f. Ammonia

   g. Chlorine

   h. Hydrochloric acid

   i. Fluorides

   j. Ozone

   k. Oxidants

   l. Photochemical smog (PAN)

   m. Arsenic [114]

   n. Mercury

   o. Ethylene

   p. Formaldehyde

   q. Ozonated hexene

   r. Organic compounds from automobile exhaust

   s. Herbicides

2. Factors affecting response of vegetation to air pollutants

   a. Genetic factors

   b. Concentration - time relations

c. Environmental and growth factors

(1) climatic factors
   (a) duration of light - photoperiod
   (b) light intensity
   (c) light quality
   (d) temperature
   (e) humidity
   (f) carbon dioxide
   (g) other climatic variations
(2) soil factors
   (a) soil moisture
   (b) soil nutrition
   (c) other soil factors
(3) growth stage
(4) other factors
   (a) synergism
   (b) antagonergism

3. Adverse effects on vegetation

a. Appearance

b. Discoloration

c. Disease

d. Growth inhibition and malformation

e. Insect infestation

f. Yield reduction

H. Effects of air pollutants on inanimate objects [36,91,113,130]

1. Pollutant found in the atmosphere with concentration sufficient to cause adverse effects on inanimate objects

a. Sulfur dioxide

b. Sulfuric acid mist

c. Nitrogen dioxide

d. Hydrogen sulfide

e. Ozone

f. Oxidants

g. Particulates and other acid mists

2. Mechanisms of deterioration in polluted atmosphere

   a. Abrasion

   b. Deposition and removal

   c. Direct chemical attack

   d. Indirect chemical attack

   e. Electrochemical corrosion

3. Factors that influence atmospheric deterioration

   a. Moisture

   b. Temperature

   c. Sunlight

   d. Air movement

   e. Other factors

      (1) angle of inclination of surface
      (2) protective measures

4. Materials vulnerable to atmospheric deterioration

   a. Metals [47,92]

      (1) ferrous
      (2) non-ferrous

   b. Building materials

   c. Paint

   d. Leather

   e. Paper

   f. Textiles

   g. Dyes

   h. Rubber and elastomers

   i. Glass and ceramics

   j. Others

5. Material damage

    a. Tarnishing and fading

    b. Discoloration

    c. Embrittlement

    d. Weakening

    e. Cracking

    f. Reducing conductivity

Air Pollution

Bibliography

1. Admur, M.O. "Air Pollution and Human Health -- Acute Biological Effects." New England Journal of Medicine. 266 (Feb. 15, 1962). pp. 348-349.

2. Admur, M.O. "Air Pollution and Human Health -- Chronic Biological Effects." New England Journal of Medicine. 266 (March 15, 1962). pp. 555-556.

3. Air Pollutant Emission Factors. Washington Operations, TRW Systems Group, McLean, Va. Apr. 1970. 332 pp.

4. Air Pollutant Emission Factors, Supplement. Washington Operations, TRW Systems Group, McLean Va. Aug. 1970. 60 pp.

5. Air Pollution. WHO Monograph #46, Geneva. 1961. N.Y.: Columbia Press. 442 pp.

6. Air Pollution Aspects of the Iron Foundry Industry. Kearney (A.T.) & Co., Chicago, Ill. Feb 1971. 260 pp.

7. Air Quality Criteria for Carbon Monoxide. National Air Pollution Control Administration, Wash., D.C. Mar. 1970. 179 pp.

8. Air Quality Criteria for Hydrocarbons. National Air Pollution Control Administration, Wash., D.C. Mar. 1970. 118 pp.

9. Air Quality Criteria for Nitrogen Oxides. Air Pollution Control Office, Environmental Protection Agency, Wash., D.C. Jan. 1971. 181 pp.

10. Air Quality Criteria for Particulate Matter. National Air Pollution Control Administration, Wash., D.C. Jan. 1969. 219 pp.

11. Air Quality Criteria for Photochemical Oxidants. National Air Pollution Control Administration, Wash., D.C. Mar. 1970. 202 pp.

12. Air Quality Criteria for Sulfur Oxides. National Air Pollution Control Administration, Wash., D.C. Jan. 1969. 186 pp.

13. American Association for the Advancement of Science. Air Conservation. AAAS Publication No. 80. 1965.

14. "An Epidemiological Appraisal of the effects of Ambient Air on Health; Particulates and Oxides of Sulfur." Journal of the Air Pollution Control Association, 19 (Sept. 1969) pp. 641-55.

15. A Study of Exhaust Emissions from Reciprocating Aircraft Power Plants. Scott Research Labs., Inc., Plumsteadville, Pa. Dec. 1970. 87 pp.

16. Athanassiadis, Yanis C. Air Pollution Aspects of Cadmium and its Compounds. Environmental Systems Div., Litton Systems, Inc., Bethesda, Md. Sept. 1969. 92 pp.

17. Athanassiadis, Yanis C. Air Pollution Aspects of Phosphorus and its Compounds. Environmental Systems Div., Litton Systems, Inc., Bethesda, Md. Sept. 1969. 86 pp.

18. Athanassiadis, Yanis C. Air Pollution Aspects of Vanadium and its Compounds. Environmental Systems Div., Litton Systems, Inc., Bethesda, Md. Sept. 1969. 105 pp.

19. Athanassiadis, Yanis C. Air Pollution Aspects of Zinc and its Compounds. Environmental Systems Div., Litton Systems, Inc., Bethesda, Md. Sept. 1969. 90 pp.

20. Atmospheric Emissions from Chlor-alkali Manufacture. Air Pollution Control Office, Environmental Protection Agency, Research Triangle Park, N.C. Jan. 1971. 116 pp.

21. Atmospheric Emissions from Nitric Acid Manufacturing Processes. National Center for Air Pollution Control, Cincinnati, Ohio. 1966. 97 pp.

22. Atmospheric Emissions from Petroleum Refineries. A Guide for Measurement and Control. Div. of Air Pollution, Public Health Service, Cincinnati, Ohio. 1960. 64 pp.

23. Atmospheric Emissions from Wet-Process Phosphoric Acid Manufacture. National Air Pollution Control Administration, Raleigh, N.C. Apr. 1970. 98 pp.

24. Bach, Wilfrid. Atmospheric Pollution. McGraw-Hill Book Co., N.Y. 1972. 144 pp.

25. Background Information -- Proposed National Emission Standards for Hazardous Air Pollutants; Asbestos, Beryllium, Mercury. Office of Air Programs, Environmental Protection Agency, Research Triangle Park, N.C. Dec. 1971. 30 pp.

26. Basis for Establishing Guides for Short-Term Explosives of the Public to Air Pollutants. Committee on Toxicology, National Academy of Sciences, National Research Council, Wash., D.C. May 1971. 16 pp.

27. Bean, Samuel L., and Howard Wall Jr. Atmospheric Emissions from Hydrochloric Acid Manufacturing Processes. National Air Pollution Control Administration, Durham, N.C. Sept. 1969. 66 pp.

28. Biologic Effects of Atmospheric Pollutants. Fluorides. Div. of Medical Sciences, National Academy of Sciences -- National Research Council, Wash., D.C. 1971. 306 pp.

29. Bond, Richard G., Conrad P. Straub, and Richard Prober (eds.) Handbook of Env. Control. Vol 1: Air Pollution. Chemical Rubber Co., Cleveland, Ohio. 1972. xii + 576 pp.

30. Bryson, R.A., and J.E. Kutzbach. Air Pollution. Commission on College Geography, Resource Paper #2. Association of American Geographers, Wash., D.C. 1968. 42 pp.

31. Buell, P., J.E. Dunn, and L. Breslow. "Cancer of the Lung and Los Angeles type Air Pollution." Cancer, 20 (Dec., 1967) 2139-2147.

32. California Div. of Highways. Air Quality Manual, Vol. VIII. Synthesis of Information of Highway Transportation and Air Quality. California State Div. of Highways, Materials and Research Dept. Dec. 1972. 41 pp.

33. Carpenter, Benjamin; W. Kenneth Poole, and Donald W. Jackson. Prevalence of Chronic Respiratory Disease in Chattanooga: Effects of Community Exposure to Nitrogen Oxides. Statistics Research Div., Research Triangle Inst., Durham, N.C. Jun. 1971. 62 pp.

34. Catcott, D.J. "Effects of Air Pollution on Animals." WHO Monograph Series No. 46. Geneva: 1961.

35. Crim, J.A., and W.D. Snowden. Asphaltic Concrete Plants. Atmospheric Emission Study. Valentine, Fisher and Tomlinson, Seattle, Wash. Nov. 1971. 93 pp.

36. Crocker, Thomas D. Urban Air Pollution Damage Functions: Theory and Measurement. Dept. of Economics, Calif. Univ., Riverside, Calif. 1970. 116 pp.

37. Cropp, G.J.A. "Effects of Air Pollution on Health," Journal Environmental Health, May-June 1973, v35, no. 6.

38. Cuffe, Stanley T. and Earl S. Schwartz. Air Pollution Aspects of Brass and Bronze Smelting and Refining Industry. National Air Pollution Control Administration, Raleigh, N.C. Nov. 1969. 65 pp.

39. Cuffe, Stanley T., and Carlton M. Dean. Atmospheric Emissions from Sulfuric Acid Manufacturing Processes. Div. of Air Pollution, Public Health Service, Cincinnati, Ohio. 1965. 136 pp.

40. Cuffe, Stanley T., and Richard W. Gerstle. Emissions from Coal-Fired Power Plants: A Comprehensive Summary. National Center for Air Pollution Control, Cincinnati, Ohio. 1967. 33 pp.

41. Durocher, Norman L. Air Pollution Aspects of Beryllium and its Compounds. Environmental Systems Div., Litton Systems, Inc., Bethesda, Md. Sept. 1969. 92 pp.

42. Durocher, Norman L. Air Pollution Aspects of Boron and its Compounds. Environmental Systems Div., Litton Systems, Inc., Bethesda, Md. Sept. 1969. 55 pp.

43. Ehrlich, R. "Effects of Air Pollutants on Respiratory Infection." Archives of Environmental Health. 6 (1963). 638.

44. Engel, R. E., D. I. Hammer, R. J. M. Horton, N. M. Lane, and L. A. Plumlee. Environmental Lead and Public Health. Air Pollution Control Office, Environmental Protection Agency, Research Triangle Park, N.C. Mar. 1971. 39 pp.

45. Esposito, J. C. Vanishing Air; the Ralph Nader Study Group Report on Air Pollution. N.Y.: Grossman. 1970. 328 pp.

46. European Conference on Air Pollution. Council of Europe, European Conference on Air Pollution at Strasburg. 24th June -- 1st July 1964. 67 pp.

47. Fink, F. W., F. H. Buttner, and W. K. Boyd. Technical-Economic Evaluation of Air-Pollution Corrosion Costs on Metals in the U.S. Columbus Labs., Battelle Memorial Inst., Columbus, Ohio. Feb. 1971. 160 pp.

48. Finkelstein, Harold. Air Pollution Aspects of Aeroallergens (Pollens). Environmental Systems Div., Litton Systems, Inc., Bethesda, Md. Sept. 1969. 118 p.

49. Finkelstein, Harold. Air Pollution Aspects of Biological Aerosols (Microorganisms). Environmental Systems Div., Litton Systems, Inc., Bethesda, Md. Sept. 1969. 109 p.

50. Finkelstein, Harold. Air Pollution Aspects of Pesticides. Environmental Systems Div., Litton Systems, Inc., Bethesda, Md. Sept. 1969. 186 p.

51. Friberg, Lars, Magnus Piscator, and Gunnar Nordberg. Cadmium in the Environment. A Toxicological and Epidemiological Appraisal. Dept. of Environmental Hygiene, Karolinska Institute, Stockholm, Sweden. April 1971. 353 p.

52. Glossary of Terms Frequently Used in Air Pollution. American Meteorological Society, Boston, Mass.

53. Goodwin, Don R., and Fred G. Rolater. Atmospheric Emissions from Thermal-Process Phospheric Acid Manufacture. National Air Pollution Control Administration, Durham, N.C. Oct. 1968. 72 p.

54. Guides for Short-term Exposures of the Public to Air Pollutants. I. Guide for Oxides of Nitrogen. Committee on Toxicology, National Academy of Sciences - National Research Council, Washington, D.C. April 1971. 33 p.

55. Guides for Short-term Exposures of the Public to Air Pollutants. II. Guide for Hydrogen Chloride. Committee on Toxicology, National Academy of Sciences - National Research Council, Washington, D.C. Aug. 1971, 16p.

56. Guides for Short-term Exposure of the Public to Air Pollutants. III. Guide for Gaseous Hydrogen Fluoride. Committee on Toxicology, National Academy of Sciences - National Research Council, Washington, D.C. Aug. 1971. 16p.

57. Hindawi, Ibrahim Joseph. Air Pollution Injury to Vegetation. National Air Pollution Control Administration, Raleigh, N.C. 1970. 49p.

58. Ipsen, J. "Episodic Morbidity and Mortality in Relation to Air Pollution" Environmental Research, 2 (Feb. 1969) pp. 137-41.

59. Kendall, David A. and Thomas Lindvall. Evaluation of Community Odor Exposure. Little (Arthur D.) Inc., Cambridge, Mass. Apr. 1971. 38 pp.

60. Kenline, Paul A. and Jeremy M. Hales. Air Pollution and the Kraft Pulping Industry. An Annotated Bibliography. Robert A. Taft, Sanitary Engineering Center, Cincinnati, Ohio. Nov. 1963. 124pp.

61. Kreichelt, Thomas E. Air Pollution Aspects of Tepee Burners Used for Disposal of Municipal Refuse. Div. of Air Pollution, Public Health Service, Cincinnati, Ohio. Sept. 1966. 39 pp.

62. Kreichelt, Thomas E., Douglas A. Kemnitz, and Stanley T. Cuffe. Atmospheric Emissions from the Manufacture of Portland Cement. National Center for Air Pollution Control, Cincinnati, Ohio. 1967. 53 pp.

63. La Belle, C. W.; Long, J. E.; and Christofano, E. E. "Synergistic Effects of Aerosols. Particulates as Carriers of Toxic Vapors," Archives of Industrial Health, 11 (1955), 297.

64. Lave, Lester B., and Seskin, Eugene P. "Air Pollution and Human Health," Science, 169 (Aug. 21, 1970), 723-33.

65. Magee, E. M.; Hall, H. J.; and Varga, G. M. Jr. Potential Pollutants in Fossil Fuels. Esso Research & Engineering Co., Linden, N.J. Government Research Lab. June 1973. 292 pp.

66. Magill, P. L.; Holden, F. R.; and Achley, C. Air Pollution Handbook. N.Y.: McGraw Hill. 1956.

67. McCormac, B. M. (ed.) Introduction to the Scientific Study of Atmospheric Pollution. D. Reidel Publishing Co., Dordrecht-Holland. 1971.

68. Millecan, Arthur A. A Survey and Assessment of Air Pollution Damage to California Vegetation in 1970. Bureau of Plant Pathology, Calif. State Dept. of Agriculture, Sacramento, Calif. June 1971. 57 pp.

69. Miner, Sydney. Air Pollution Aspects of Ammonia. Environmental Systems Div., Litton Systems, Inc., Bethesda, Md. Sept. 1969. 51 pp.

70. Miner, Sydney. Air Pollution Aspects of Barium and its Compounds. Environmental Systems Div., Litton Systems, Inc., Bethesda, Md. Sept. 1969. 69 pp.

71. Miner, Sydney. Air Pollution Aspects of Hydrogen Sulfide. Environmental Systems Div., Litton Systems, Inc., Bethesda, Md. Sept. 1969. 107 pp.

72. National Industrial Pollution Control Council. Air Pollution by Sulfur Oxides. Staff Report. Feb. 1971. 24 pp.

73. NIPCC. Exhaust Emissions from Gas Turbine, Aircraft Engines. Sub-Council Report. Feb. 1971. 28 pp.

74. National Inventory of Sources and Emissions: Cadmium, Nickel, and Asbestos - 1968. Cadmium, Section I. David (W.E.) and Associates, Leawood, Kansas. Feb. 1970. 53 pp.

75. National Inventory of Sources and Emissions: Cadmium, Nickel, and Asbestos - 1968. Nickel, Section II. David (W.E.) and Associates, Leawood, Kansas. Feb. 1970. 46 pp.

76. National Inventory of Sources and Emissions: Cadmium, Nickel, and Asbestos - 1968. Asbestos, Section III. David (W.E.) and Associates, Leawood, Kansas. Feb. 1970. 56 pp.

77. National Inventory of Sources and Emissions. Barium, Boron, Copper, Selenium, and Zinc, 1969 -- Section I, Barium. David & Associates, Leawood, Kansas. May 1972. 56 pp.

78. National Inventory of Sources and Emissions. Barium, Boron, Copper, Selenium, and Zinc, 1969 -- Section II, Boron. David & Associates, Leawood, Kansas. June 1972. 51 pp.

79. National Inventory of Sources and Emissions. Barium, Boron, Copper, Selenium, and Zinc, 1969 -- Section III, Copper. David & Associates, Leawood, Kansas. April 1972. 74 pp.

80. National Inventory of Sources and Emissions. Barium, Boron, Copper, Selenium, and Zinc, 1969 -- Section IV, Selenium. David & Associates, Leawood, Kansas. April 1972. 57 pp.

81. National Inventory of Sources and Emissions. Barium, Boron, Copper, Selenium, and Zinc, 1969 -- Section V, Zinc. David & Associates, Leawood, Kansas. May 1972. 85 pp.

82. National Survey of the Odor Problem. Phase I of a Study of the Social and Economic Impact of Odors. Copley International Corp., N.Y. Jan. 1970. 258 pp.

83. National Survey of the Odor Problem. Phase I of a Study of the Social and Economic Impact of Odors. Appendix. Copley International Corp., N.Y. Jan. 1970. 360 pp.

84. National Survey of the Odor Problem. Phase II of a Study of the Social and Economic Impact of Odors. Copley International Corp. N.Y. Nov. 1971. 313 pp.

85. Nationwide Inventory of Air Pollutant Emissions - 1968. Div. of Air Quality and Emission Data, National Air Pollution Control Administration, Arlington, Va. Aug. 1970. 44 pp.

86. Novick, Sheldon. Air Pollution. Virginia Brodine, Harcourt Brace, N.Y. 1973. 206 pp.

87. Olsen, Douglas A., and Haynes, James L. Air Pollution Aspects of Carcinogens. Environmental Systems Div., Litton Systems, Inc., Bethesda, Md. Sept. 1969. 131 pp.

88. Partee, Frank. Air Pollution in the Coffee Roasting Industry. Public Health Service, Cincinnati, Ohio, Div. of Air Pollution. Sept. 1964. 20 pp.

89. Platt, M.; Baker, R. C.; Bastress, E. K.; Chng, K. M.; and Siegel, R. D. The Potential Impact of Aircraft Emission upon Air Quality. Northern Research & Engineering Corp., Cambridge, Mass. Dec. 1971. 330 pp.

90. Requeriro, Jose F. Collection & Assessment of Aircraft Emissions. Teledyne Continental Motors, Muskegon, Mich. Oct. 1971. 130 pp.

91. Ridker, R. G. Economic Costs of Air Pollution. N.Y.: Praeger. 1967. 214 pp.

92. Robbins, Robert C. Inquiry into the Economic Effects of Air Pollution on Electrical Contacts. Stanford Research Inst., Menlo Park, Calif. Apr. 1970. 101 pp.

93. Robert, J. Lillie. Air Pollutants Affecting the Performance of Domestic Animals--A Literature Review. US Dept. of Agriculture. Report 380. Jan. 1972. 109 pp.

94. Robinson, George D. "Long-term Effects of Air Pollution," Instruments and Control Systems. June 1972. v. 45. n. 6.

95. Robinson, G. D. Long-term Effects on Air Pollution--A Survey. Center for the Environment and Man, Inc., Hartford, Conn. June 1970. 50 pp.

96. Ross, R. D. (ed.) Air Pollution & Industry. Van Nostrand Reinhold Co., N.Y. 1972. 489 pp.

97. Rutgers, Dept. of Biology. 1971 Survey & Assessment of Air Pollution Damage to Vegetation in New Jersey. Rutgers State U., Dept. of Biology. Oct. 1972.

98. Schueneman, Jean J.; High, M. D.; and Bye, W. E. Air Pollution Aspects of the Iron & Steel Industry. Robert A. Taft Sanitary Engineering Center, Cincinnati, Ohio. June 1963. 128 pp.

99. Shannon, L. J.; Gorman, P. G.; and Retchel, M. Particulate Pollutant System Study. Vol. II. Fine Particle Emissions. Midwest Research Inst., Kansas City, Mo. Aug. 1971. 343 pp.

100. Singer, S. F. (ed.) Global Effects of Atmospheric Pollution. Reidel Publishing Co., Dorcrecht, Netherlands.

101. Smith, Walter S. Atmospheric Emissions from Fuel Oil Combustion; An Inventory Guide. Robert A. Taft Sanitary Engineering Center, Cincinnati, Ohio. Nov. 1962. 103 pp.

102. Smith, W. S., and Gruber, C. W. Atmospheric Emissions from Coal Combustion, An Inventory Guide. Div. of Air Pollution, Public Health Service, Cincinnati, Ohio. Apr. 1966. 114 pp.

103. Stahl, Quade R. Air Pollution Aspects of Aldehydes. Environmental Systems Div., Litton Systems, Inc., Bethesda, Md. Sept. 1969. 149 pp.

104. Stahl, Quade R. Air Pollution Aspects of Chlorine Gas. Environmental Systems Div., Litton Systems, Inc., Bethesda, Md. Sept. 1969. 90 pp.

105. Stahl, Quade R. Air Pollution Aspects of Ethylene. Environmental Systems Div., Litton Systems, Inc., Bethesda, Md. Sept. 1969. 65 pp.

106. Stahl, Quade R. Air Pollution Aspects of Hydrochloric Acid. Environmental Systems Div., Litton Systems, Inc., Bethesda, Md. Sept. 1969. 82 pp.

107. Stahl, Quade R. Air Pollution Aspects of Mercury and its Compounds. Environmental Systems Div., Litton Systems, Inc., Bethesda, Md. Sept. 1969. 108 pp.

108. Stahl, Quade R. Air Pollution Aspects of Selenium and its Compounds. Environmental Systems Div., Litton Systems, Inc., Bethesda, Md. Sept. 1969. 88 pp.

109. Sterling, T. D., et al. "Urban Morbidity and Air Pollution." Archives of Environmental Health. Aug. 13, 1966. pp. 158-70.

110. Stern, A. C. (ed.) Air Pollution. v. 1, 2, & 3. Academic Press, N.Y. 1968.

111. Study of Air Pollution Aspects of Various Roadway Configurations. Re-entry and Environmental Systems Div., General Electric Co., Philadelphia, Pa. Sept. 1971. 211 pp.

112. Study of Jet Aircraft Emissions and Air Quality in the Vicinity of the Los Angeles International Airport. Los Angeles County Air Pollution Control District, Calif. Apr. 1971. 190 pp.

113. Study to Determine Residential Soiling Costs of Particulate Air Pollution. Booz-Allen & Hamilton, Inc., Wash., D.C. Oct. 1970. 201 pp.

114. Sullivan, Ralph J. Air Pollution Aspects of Arsenic and Its Compounds. Environmental Systems Div., Litton Systems, Inc., Bethesda, Md. Sept. 1969. 72 pp.

115. Sullivan, Ralph J. Air Pollution Aspects of Chromium and its Compounds. Environmental Systems Div., Litton Systems, Inc., Bethesda, Md. Sept. 1969. 86 pp.

116. Sullivan, Ralph J. Air Pollution Aspects of Iron and Its Compounds. Env. Systems Div., Litton Systems, Inc., Bethesda, Md. Sept. 1969. 106 pp.

117. Sullivan, Ralph J. Air Pollution Aspects of Manganese and Its Compounds. Env. Systems Div., Litton Systems, Inc., Bethesda, Md. Sept. 1969. 63 pp.

118. Sullivan, Ralph J. Air Pollution Aspects of Nickel and its Compounds. Environmental Systems Div., Litton Systems, Inc., Bethesda, Md. Sept. 1969. 76 pp.

119. Sullivan, Ralph J. Air Pollution Aspects of Odorous Compounds. Environmental Systems Div., Litton Systems, Inc., Bethesda, Md. Sept. 1969. 258 pp.

120. Sullivan, Ralph J., and Athanassiadis, Yanis C. Air Pollution Aspects of Abestos. Environmental Systems Div., Litton Systems, Inc., Bethesda, Md. Sept. 1969. 105 pp.

121. Systems Analysis of Emissions and Emissions Control in the Iron Foundry Industry. Vol. I. Text. Kearney (A. T.) and Co., Chicago, Ill. Feb. 1971. 368 pp.

122. Systems Analysis of Emissions and Emissions Control in the Iron Foundry Industry. Vol. II. Exhibits. Kearney (A. T.) and Co., Chicago, Ill. Feb. 1971. 175 pp.

123. Systems Analysis of Emissions and Emissions Control in the Iron Foundry Industry. Vol. III. Appendix. Kearney (A. T.) and Co., Chicago, Ill. Feb. 1971. 346 pp.

124. United States Environmental Protection Agency. *Compilation of Air Pollutant Emission Factors* (2nd ed.) United States Environmental Protection Agency, Office of Air and Water Programs, Office of Air Quality Planning and Standards, Research Triangle Park, N.C. Apr. 1973.

125. United States Environmental Protection Agency. *Environmental Lead and Public Health*. By Engel, R. E.; Hammer, D. I.; Horton, R. J. M.; Lane, N. M.; Plumlee, L. A. United States Environmental Protection Agency, Air Pollution Control Office, Research Triangle Park, N.C. Mar. 1971. 34 pp.

126. United States Environmental Protection Agency. *Hydrochloric Acid and Air Pollution*. Office of Air Programs. Publication: AP-100. July 1971.

127. United States Office of Management and Budget. *A Reivew of the Health Effects of Sulfur Oxides*. White House Sulfur Oxide Study.

128. Vandegrift, A. E.; Shannon, L. J.; Gorman, P. G.; Lawless, E. W.; and Sallee, E. E. *Particulate Pollutant System Study. Vol. I. Mass Emissions*. Midwest Research Inst., Kansas City, Mo. May 1971. 376 pp.

129. Vandegrift, A. E., *et al*. *Particulate Pollutant System Study. Vol. III. Handbook of Emission Properties*. Midwest Research Inst., Kansas City, Mo. May 1971. 613 pp.

130. Vars, R. Charles, and Sorenson, Gary W. *Study of the Social and Economic Effects of Changes in Air Quality*. Water Resources Research Inst., Oregon State U., Corvallis, Oregon, June 1971. 70 pp.

131. Voelz, F. L., *et al*. "Automotive Exhaust Emission Levels by Geographic Area and Vehicle Make." *APCA J*. Dec. 1972. v. 22. n. 12.

132. Williams, J. D.; Farmer, J. R.; Stephenson, R. B.; Evans, G. G.; and Dalton, R. B. *Air Pollutant Emissions Related to Land Area--A Basis for a Preventive Air Pollution Control Program*. National Air Pollution Control Administration, Durham, N.C. July 1968. 19 pp.

133. Wolkonsky, Peter Malia. "Pulmonary Effects of Air Pollution." *Archives of Environmental Health*. Oct. 19, 1969. pp. 586-92.

General Bibliographical Sources on Air Pollution

Air Pollution Aspects of Emission Sources: Cement Manufacturing. A Bibliography with Abstracts. Office of Air Programs, Env. Protection Agency, Research Triangle Park, N.C. May 1971. 51 pp.

Air Pollution Aspects of Emission Sources: Electric Power Production--A Bibliography with Abstracts. Office of Air Programs, EPA, Research Triangle Park, N.C. May 1971. 319 pp.

Air Pollution Aspects of Emission Sources: Iron and Steel Mills. A Bibliography with Abstracts. Office of Air Programs, EPA, Research Triangle Park, N.C. May 1972. 86 pp.

Air Pollution Aspects of Emission Sources: Municipal Incineration. A Bibliography with Abstracts. Air Pollution Control Office, EPA, Research Triangle Park, N.C. May 1971. 101 pp.

Air Pollution Aspects of Emission Sources: Sulfuric Acid Manufacturing. A Bibliography with Abstracts. Office of Air Programs, EPA, Research Triangle Park, N.C. May 1971. 64 pp.

Air Pollution Translations: A Bibliography with Abstracts. Vol. I. National Air Pollution Control Administration, Arlington, Va. May 1969. 173 pp.

Air Pollution Translations: A Bibliography with Abstracts. Vol. II. National Air Pollution Control Administration, Arlington, Va. Apr. 1970. 115 pp.

Campbell, Irene R., and Mergard, Estelle G. Biological Aspects of Lead: An Annotated Bibliography. Part I. Literature from 1950 through 1964. Kettering Lab., Cincinnati Univ., Ohio. May 1972. 568 pp.

Campbell, Irene R., and Mergard, Estelle G. Biological Aspects of Lead: An Annotated Bibliography. Part II. Literature from 1950 through 1964. Kettering Lab., Cincinnati Univ., Ohio. May 1972. 380 pp.

Asbestos and Air Pollution: An Annotated Bibliography. Air Pollution Control Office, EPA, Research Triangle Park, N.C. Feb. 1971. 105 pp.

Beryllium and Air Pollution: An Annotated Bibliography. Air Pollution Control Office, EPA, Research Triangle Park, N.C. Feb. 1971. 74 pp.

Chlorine and Air Pollution: An Annotated Bibliography. Office of Air Programs, EPA, Research Triangle Park, N.C. July 1971. 108 pp.

Hydrocarbons and Air Pollution: An Annotated Bibliography. Part I. Categories A to E, and Part II. Categories F to M and Indexes. National Air Pollution Control Administration, Raleigh, N.C. Oct. 1970. 1184 pp.

Hydrochloric Acid and Air Pollution: An Annotated Bibliography. Office of Air Programs, EPA, Research Triangle Park, N.C. July 1971. 114 pp.

Nitrogen Oxides: An Annotated Bibliography. National Air Pollution Control Administration, Raleigh N.C. Aug. 1970. 636 pp.

Nuttonson, M.Y. AICE Survey of USSR Air Pollution Literature. Vol. VII. Measurements of Dispersal and Concentration, Identification, and Sanitary Evaluation of Various Air Pollutants, with Special Reference to the Environs of Electric Power Plants and Ferrous Metallurgical Plants. 1971. 112 pp.

Nuttonson, M.Y. AICE Survey of USSR Air Pollution Literature. Vol. X. The Toxic Components of Automobile Exhaust Gases: Their Composition Under Different Operating Conditions, and Methods of Reducing their Emission. 1971. 139 pp.

Nuttonson, M.Y. AICE Survey of USSR Air Pollution Literature. Vol. VIII. A Compilation of Technical Reports on the Biological Effects and the Public Health Aspects of Atmospheric Pollutants. July 1971. 170 pp.

Nuttonson, M.Y. AICE Survey of USSR Air Pollution Literature. Vol. XI. A Second Compilation of Technical Reports on the Biological Effects and the Public Health Aspects of Atmospheric Pollutants. Jan. 1972. 154 pp.

Nuttonson, M.Y. AICE Survey of USSR Air Pollution Literature. Vol. XV. A Third Compilation of Technical Reports on the Biological Effects and the Public Health Aspects of Atmospheric Pollutants. July 1972. 154 pp.

Nuttonson, M.Y. AICE Survey of USSR Air Pollution Literature. Vol. II. Effects and Symptoms of Air Pollutes on Vegetation, Resistance and Susceptibility of Different Plane Species in Various Habitats, in Relation to Plant Utilization for Shelter Belts and as Biological Indicators. Dec. 1969. 108 pp.

Nuttonson, M.Y. AICE Survey of USSR Air Pollution Literature. Vol. III. The Susceptibility or Resistance to Gas and Smoke of Various Arboreal Species Grown under Diverse Environmental Conditions in a Number of Industrial Regions of the Soviet Union. Dec. 1969. 126 pp.

Nuttonson, M.Y. AICE Survey of USSR Air Pollution Literature. Vol. IX. Gas Resistance of Plants with Special Reference to Plant Biochemistry and to the Effects of Mineral Nutrition. American Inst. of Crop Ecology, Silver Spring, Md. Jan. 1971. 121 pp.

United States Environmental Protection Agency. Odors and Air Pollution: A Bibliography with Abstracts. Air Pollution Technical Information Center, Office of Air Programs, Research Triangle Park, N.C. Oct. 1972.

III. Solid Wastes [1,2,3,5,6,7]

A. Sources of Solid Wastes

1. Urban

   a. Residential, domestic
   b. Private institutions [4]
   c. Offices and public facilities
   d. Private health services
   e. Commercial
   f. Streets, sidewalks, alleys, parking lots, vacant lots

2. Agricultural

3. Construction and demolition

4. Industrial

   a. Extractive industry
   b. Food and kindred products
   c. Tobacco industry
   d. Ordnance and accessories
   e. Textile and apparel
   f. Lumber and wood products
   g. Furniture and fixtures
   h. Paper and allied products
   i. Printing, publishing, and allied products
   j. Chemicals and allied products
   k. Plumbing, heating, air conditioning, and special trade contractors
   l. Petroleum refining
   m. Rubber and plastic products
   n. Leather and allied products
   o. Stone, clay, and glass
   p. Primary metals
   q. Fabricated metals
   r. Nonelectrical machinery
   s. Electrical machinery
   t. Transportation equipment
   u. Professional and scientific instruments
   v. Miscellaneous manufacturing

B. Composition of Solid Wastes

1. Ashes, residue, slag, grit
2. Debris, dirt, masonry, asphaltic material
3. Garbage
4. Dead animal waste
5. Manure
6. Garden wastes
7. Paper products

8. Wood
9. Glass and ceramics
10. Cloth, rag, fiber
11. Plastics
12. Rubber
13. Leather
14. Metals
    - Ferrous
    - Nonferrous
15. Oil, paint, chemicals
16. Explosives
17. Radioactive wastes
18. Pathologic wastes
19. Others

C. Important Characteristics of Solid Wastes

1. Bulkiness
2. Combustibility
3. Rats, flies, and other disease vectors accomodating (providing food and harborage for disease vectors)
4. Solubility

D. Transport and the Transformation of Solid Wastes

1. Chemical reaction
2. Biological stabilization
3. Physical action
    a. Leaching
    b. Washout
    c. Vaporization

E. Effects of Solid Wastes

1. Health hazards
2. Aesthetics
3. Odor nuisance
4. Occupational hazard

Solid Wastes

Bibliography

1. Bond, Richard G., and Straub, Conrad P. Handbook of Environmental Control Vol. II: Solid Waste. CRC Press, The Chemical Rubber Co., Cleveland, Ohio. 1973. 580 p.

2. Golueke, C. G., and McGauhey, P. H. Comprehensive Studies of Solid Wastes Management. Sanitary Engineering Research Laboratory, Univ. of Calif. 1969.

3. Heslep, John. Solid Waste. C.O.A.P. April 1971.

4. Singer, Rexford D.; DuChene, Alain G.; and Vick, Nichole J. Hospital Solid Waste. School of Public Health, Minnesota University, Minneapolis. Mar. 1974. 205 p.

5. Solid Waste: A New Natural Resource. Dept. of Chemical Engineering, West Virginia Univ., Morgantown. May 1971. 18 p.

6. Solid Waste Management. Prepared by Ad Hoc Group for the Office of Science and Technology, Executive Office of the President, Washington D.C. May 1969. 111 p.

7. U.S. Dept. of Health, Education, and Welfare. A Systems Study of Solid Waste Management in the Fresno Area. 1969.

General Bibliographical Source on Solid Wastes

U.S. Environmental Protection Agency. Solid Waste Management: a list of available literature (report SW-58.16). U.S. Government Printing Office. Oct. 1972. 45 p.

IV. Noise [7,12,17,18,30,34,35,38,44,45,49,50]

A. Sources of Noise [11]

1. Domestic - use of appliances [31]

2. Construction [31]

3. Manufacturing [20,22,32]

4. Power generation

5. Transportation [9,43,46]

a. Aircraft, SST [14,16,21,26,37,40,42]

b. Other surface transportation [48]

B. Parameters of Importance in Determining the Effects of Noise

1. General

a. Characteristics of noise [29]

(1) amplitude - loudness
(2) frequency - pitch
(3) duration - time
(a) ongoing
(b) impulsive

b. Variety of noise

c. Location of noise exposure

d. Time of noise exposure at any given location

2. Annoyance factors

a. Noise climate or background noise against which a particular noise occurs

b. Previous experience of the individual or community with the particular noise

c. Attitude of the individual or community regarding the contribution of the activities associated with the noise source to the general well-being of the community

d. Fear associated with activities of noise source

e. Extent to which individual or community believes the noise source could be controlled

f. Information content of noise

g. Degree of interference with activity

h. Socioeconomic status and educational level of the individual or community

C. Effects on Man

1. Adverse effects on man [3,4,5,6,8,24,33]

a. Health effect [1,10,47]

(1) impairment of hearing mechanism [2,15,36,51]
(2) pain and other physiological response [19]
(3) mental health [23]

b. Interference

(1) masking of communication
(2) interference with sleep
(3) effect on performance [13,39]
(4) annoyance

2. Nature of effects

a. Temporary

b. Permanent

D. Effects on Animals [25]

1. Impairment of hearing mechanism

2. Masking of communication

3. Behavioral changes

4. Physiological changes

E. Effects on Inanimate Objects [28]

1. Objects vulnerable to damage

a. Secondary structural elements, e.g. window, plaster

b. Aged, wearing, and unstable structure

c. Ductile metal

2. Types of damage

a. Failure of structure

b. Weakening of materials - e.g. sonic fatigue

Noise

Bibliography

1. Alexander, Walter. "Some Harmful Effects of Noise." Canadian Medical Association Journal. 99. July 6, 1968. pp. 27-31.

2. American Academy of Opthamology and Otolaryngology, Committee on Conservation of Hearing, Subcommittee on Noise. Transactions of the American Academy of Opthalmology and Otolaryngology. 68 (1964). Supplement.

3. Baren, R. A. "Noise and Urban Man." American Journal of Public Health. 58 (November, 1968). pp. 2060-66.

4. Bell, A. Noise, An Occupational Hazard and Public Nuisance. Public Health Paper #30, World Health Organization, Geneva, Switzerland. 1966.

5. Broadbent, Donald E. "Effects of Noise on Behavior." Handbook of Noise Control. Edited by Cyril M. Harris. New York: McGraw-Hill Book Company, Inc. 1957. 10-1 -- 10-33.

6. Broadbent, D. E. "Non-Auditory Effects of Noise." The Advancement of Science. (January, 1961). pp. 406-409.

7. Burns, William. Noise and Man. J. B. Lippincott Company, Philadelphia. 1973. 459 p.

8. Central Institute for the Deaf. Effects of Noise on People. EPA Document NTID 300.7.

9. Chalupnick, J. D. (ed). Transportation Noise, A Symposium on Acceptability Data. Seattle: University of Washington Press. 1970.

10. Cohen, Alexander. "Noise Effects on Health, Productivity, and Well-being." Transactions of the New York Academy of Sciences. 30. (May, 1968). pp. 910-18.

11. Community Noise. Office of Noise Abatement and Control, Environmental Protection Agency, Washington, D.C. 31 Dec., 1971. 212 p.

12. Control of Noise. National Physical Laboratories, Her Majesty's Stationery Office, London, England. 1962. 434 p.

13. Environmental Pollution: Noise Pollution -- Noise Effects on Human Performance. Defense Documentation Center, Alexandria, Va. Nov. 1973. 170 pp.

14. Environmental Pollution: Noise Pollution -- Sonic Boom. Defense Documentation Center: Alexandria, Va. Nov. 1973. 179 pp.

15. Glorig, A. Jr. Noise and Your Ear. New York and London: Grune & Stratton. 152 p.

16. Goodman, Robert F. Airport Noise and the Changing Patterns of Airport-Community Politics. University of Southern California, Center for Urban Affairs. Feb. 1973. 38 pp.

17. Harris, C. M. Handbook of Noise Control. McGraw-Hill Book Co. 1957.

18. Hearings before the Subcommittee on Air and Water Pollution of the Committee on Public Works, United States Senate, 92nd Contress. Noise Pollution. U.S. Government Printing Office. 1972. 604 pp.

19. Horn, K. "City Noises and Their Effects on the Central Nervous System." Deutsche Gesundheitswesen. 15 (Aug. 4, 1960). pp. 1617-22.

20. Industrial Noise Manual. American Industrial Hygiene Association, Noise Committee, 1966. Detroit, Michigan: American Industrial Hygiene Association. 171 p.

21. Investigation and Study of Aircraft Noise Problems. House Report No. 36, 88th Congress.

22. Karplus, H. B., and Bonvallet, G. L. "A Noise Survey of Manufacturing Industries." American Industrial Hygiene Association Quarterly. 14. pp. 235-85. 1953.

23. Kryter, Karl D. "Psychological Reactions to Aircraft Noise." Science. 151 (March, 1966). pp. 1346-55.

24. Kryter, Karl D. The Effects of Noise on Man. New York: Academic Press. 1970.

25. Memphis State University, Tennessee. Effects of Noise on Wildlife and Other Animals. U.S. Environmental Protection Agency: Wash., D.C., NTED 300.5. 1971. 74 pp.

26. National Aircraft Noise Symposium -- Jamaica, New York. Report of Proceedings. 1965.

27. National Bureau of Standards. Economic Impact of Noise. EPA Document NTID 300.14.

28. National Bureau of Standards. Effects of Sonic Boom and Other Impulsive Noise on Structures. EPA Document NTID 300.12.

29. National Bureau of Standards. Fundamentals of Noise: Measurement, Rating Schemes, and Standards. EPA Document NTID 300.15. 1971.

30. Noise Facts Digest. Informatics, Inc., Rockville, Md. June 1972. 206 p.

31. Noise from Construction Equipment and Operations, Building Equipment, and Home Appliances. Bolt Beranek and Newman, Inc., Cambridge, Mass. 31 Dec., 1971. 337 p.

32. Noise from Industrial Plants. Prepared by L. S. Goodfriend Associates under contract EPA 68-04-0044.

33. "Noise Nuisances and Their Effects, Individual and Public Health Aspects." Royal Society of Health Journal. 83 (July-Aug., 1963). pp. 218-23.

34. Noise Pollution Resources Compendium. New Mexico University, Albuquerque, Technology Application Center. 31 Mar., 1973. 129 p.

35. "Noise -- Sound Without Value." Committee on Environmental Quality, U.S. Federal Council for Science and Technology. Man's Impact on Environment. Ed. by Thomas R. Detroyler. New York: Mcgraw-Hill Company. 1971. pp. 175-189.

36. Relations of Hearing Loss to Noise Exposure. Z24-X-2 Report, American Standards Association, New York, New York. 1954.

37. Report of Generation and Propagation of Sonic Boom. National Academy of Sciences, Committee on SST - Sonic Boom, Subcommittee on Research. Oct. 1967.

38. Rodda, Michael. Noise and Society. Edinburgh: Oliver & Body, Ltd. 1967.

39. Schoenberger, R. W., et al. "Human Performance as a Function of Changes in Acoustic Noise Levels." Journal of Engineering Psychology. 4 (1965). pp. 108-19.

40. Shurcliff, W. A. SST and Sonic Boom Handbook. New York: Friends of the Earth and Ballantine Books. 1970.

41. Social Impact of Noise. National Bureau of Standards, Washington, D.C. 31 Dec., 1971. 25 p.

42. The Airport and Its Neighbors. Report of the President's Airport Commission. 1952.

43. Transportation Noise and Noise from Equipment Powered by Internal Combustion Engines. Wyle Labs., Inc., El Segundo, Calif. 31 Dec., 1971. 427 p.

44. Public Health and Welfare Criteria for Noise. Office of Noise Abatement and Control, US EPA. July 27, 1973.

45. U.S. Environmental Protection Agency. Report to the President and Congress on Noise. US EPA, Washington, D.C. Feb. 1972.

46. U.S. Office of the Secretary of Transportation. Transportation Noise and Its Control. June 1972.

47. Ward, W. D., and Fricke, J. E. Proceedings of the Conference on Noise as a Public Health Hazard. Washington: The American Speech and Hearing Association. ASHA Report 4. 384 p.

48. "Wayside Noise Levels to be Expected from Operation of the Bay Area Rapid Transit System Trains." Wilson, Ihrig and Association, Acoustical Consultants. Dec. 9, 1969.

49. Whitehorn, Norman C. Noise Pollution: Effects and Control. Texas A & M Univ., Sea Grant Program. June 1972.

50. Wilson, Sir Alan. Noise, Final Report. Committee on the Problem of Noise, CMND 2056, Her Majesty's Stationery Office, London, England. July 1963.

51. Yaffee, C. D., and Jones, H. H. Noise and Hearing -- Relationship of Industrial Noise to Hearing Acuity in a Controlled Population. U.S. Public Health Service Publication #850, GPO, Washington, D.C.

Resource Document on Noise

Noise Programs of Professional/Industrial Organizations, Universities and Colleges. Office of Noise Abatement and Control, Environmental Protection Agency, Washington, D.C. Dec. 1971. 86 p.

V. Radiation [1]

A. Sources of Radiation

1. Natural sources

a. Naturally occurring radioactive substances in the lithosphere

(1) Ra^{226}

(2) U^{238}

(3) Th^{232}

(4) K^{40}

(5) V^{50}

(6) Rb^{87}

(7) In^{115}

(8) La^{138}

(9) Sm^{147}

(10) Lu^{176}

b. Cosmic radiation of extraterrestrial origin. Products of cosmic radiation:

(1) H^{3}

(2) C^{14}

(3) Be^{7}

(4) S^{35}

(5) P^{33}

(6) P^{32}

2. Man-made source

a. Medical procedures - medical and dental X-ray examination

b. Television, microwave oven, and other appliances [2,3,5,11,13,14]

c. Nuclear power plants [7,8,9,12]

d. Nuclear weapon tests

e. Radioactive minerals mining

f. Constructive uses of nuclear explosives - excavation

g. Leakage from radioactive waste deposit

B. Radiation Hazardous to Life - Ionizing Radiation

1. Major ionizing radiation

a. Alpha radiation

b. Beta radiation

c. Gamma radiation

2. Other types of ionizing radiation that are encountered less frequently

a. Protons

b. Neutrons

c. Deutrons

d. Other nuclei of sufficiently great energy

C. Effects of Radiation on Life [2,4,6,10]

1. Biological classification of radiation effects

a. Somatic

b. Genetic

2. Factors affecting the effects of radiation

a. Quantity of release

b. Half-life of radioactive materials

c. Location of release - local, tropospheric, stratospheric, fallout

d. Biological uptake - the change and transfer of radioactive materials

3. Adverse effects on life

a. Direct effects - fragmentation of biologically important molecules such as DNA molecules in the cell nucleus

b. Indirect effects - fragmentation of biologically less vital molecules with the formation of reactive ions and free radicals that can later affect more important molecules and impair their usefulness

Radiation

Bibliography

1. 1969 Annual Report to the Congress on the Administration of the Radiation Control for Health and Safety Act of 1968. Public Law 90-602. Office of the Director, Bureau of Radiological Health, Rockville, Md. April 1970. 97 p.

2. Cleary, Stephen F. Biological Effects and Health Implications of Microwave Radiation. Bureau of Radiological Health, Rockville, Md. Sept. 1969. 269 p.

3. Electric Product Radiation and the Health Physicist. Div. of Electronic Products, Bureau of Radiological Health, Rockville, Md. Oct. 1970. 470 p.

4. Guidelines to Radiological Health. Bureau of Radiological Health, Rockville, Md. Sept. 1968. 185 p.

5. Harris, Jesse Y. Electronic Product Inventory Study. Div. of Electronic Products, Bureau of Radiological Health, Rockville, Md. Nov. 1970.

6. Leach, William M. Biological Aspects of Ultraviolet Radiation. A Review of Hazards. Div. of Biological Effects, Bureau of Radiological Health, Rockville, Md. Sept. 1970. 38 p.

7. Logsdon, Joe E., and Chissler, Robert I. Radioactive Waste Discharges to the Environment from Nuclear Power Facilities. Div. of Environmental Radiation, Bureau of Radiological Health, Rockville, Md. March 1970. 84 p.

8. Logsdon, Joe E., and Robinson, Thomas L. Radioactive Waste Discharges to the Environment from Nuclear Power Facilities. Addendum - 1. Surveillance and Inspection Div. Office of Radiation Programs, Washington, D.C. Oct. 1971. 33 p.

9. Miner, Sydney. Air Pollution Aspects of Radioactive Substances. Environmental Systems Div., Litton Systems, Inc., Bethesda, Md. Sept. 1969. 159 p.

10. Moore, Wellington Jr. Biological Aspects of Laser Radiation. A Review of Hazards. Bureau of Radiological Health, Rockville, Md. Jan. 1969. 18 p.

11. Murphy, Emmet; Stewart, Harold; Coppola, Steven; and Modine, Norman. X-Ray Emission from Shunt Regulator Tubes for Color Television Receivers. Medical and Occupational Radiation Program, National Center for Radiological Health, Rockville, Md. June 1967. 25 p.

12. Oates, William H. Jr. Radiation Exposure Overview. Nuclear Power Reactors and the Population. Office of Criteria and Standards, Bureau of Radiological Health, Rockville, Md. Jan. 1970. 32 p.

13. Rosenstein, Marvin; Brill, Warren A.; and Showalter, Charles K. Radiation Exposure Overview: Microwave Ovens and the Public. Office of Criteria and Standards, Bureau of Radiological Health, Rockville, Md. July 1969. 30 p.

14. Seabron, La Vert C., and Coopersmith, Lewis W. Results of the 1970 Microwave Oven Survey. Div. of Electronic Products, Bureau of Radiological Health, Rockville, Md. Aug. 1971. 158 p.

Resource Documents on Radiation

Publications Index, July 1972. Office of the Director, Bureau of Radiological Health, Rockville, Md. Nov. 1970. 228 p.

University Curriculums and Fellowships in Radiological Health. Office of the Director, Bureau of Radiological Health, Rockville, Md. 1970. 88 p.

VI. Weather and Climate

A. Schematic for Analysis

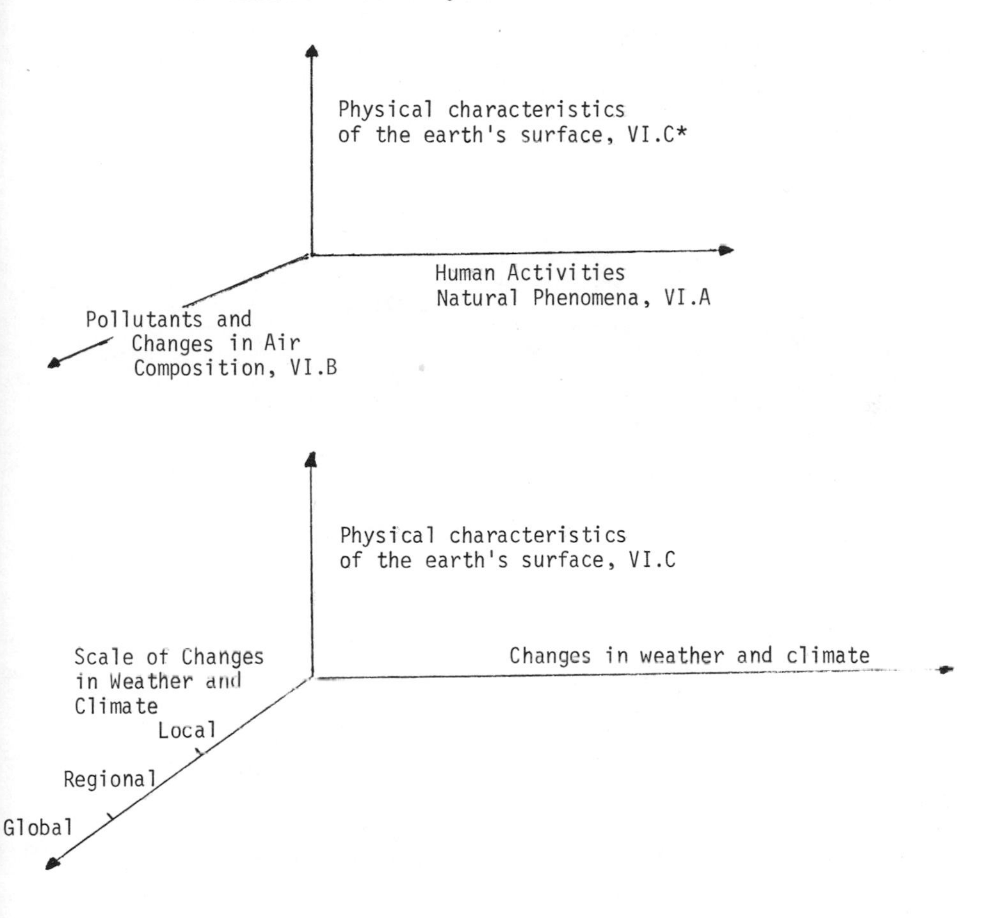

*Refers to other topics in this section

B. Natural Phenomena and Human Activity

1. Natural phenomena

a. Volcanic eruption

b. Exchange of CO_2 between atmosphere and ocean

c. Exchange of CO_2 between atmosphere and forest

2. Human activity

a. Thermal power output (combustion of fossil fuel)

b. Changes of landscape -- population growth, urbanization and overgrazing

c. Cloud formation

(1) operation of subsonic jets in the troposphere
(2) operation of supersonic transports in the stratosphere

d. Destruction of forest

C. Pollutants and Changes in Air Composition

1. Heat

2. Particulate

3. Changes in concentration of CO_2

D. Physical Characteristics of the Earth's Surface which have a Bearing on Weather and Climate[16]

1. Reflectivity

2. Heat capacity and conductivity

3. Availability of water and dust

4. Aerodynamic roughness

5. Emissivity in the infrared bend

6. Heat released to the ground

E. Changes in Weather and Climate[2,4,6,8,9]

1. Temperature - heat island effect, "greenhouse" effect [3,14]

2. Humidity

3. Cloudiness
4. Visibility [10]
5. Radiation
6. Precipitation [1,5,7]
7. Wind
    a. Speed
    b. Direction

Effects on Weather and Climate

Bibliography

1. Atkinson, B. W. "The Reality of the Urban Effect on Precipitation--a Case Study Approach," presented at W.M.O. Symposium on Urban Climates and Building Climatology. Brussels, Belgium. Oct. 1969.

2. Chandler, T. J. "London's Urban Climate." Geography Journal 127, 279. 1962.

3. Chandler, T. J. "Surface Effects of Leicester's Heat-island." E. Midland Geographers. No. 15, 32. 1961.

4. Chandler, T. J. "Wind as a Factor of Urban Temperatures--a Survey in North-east London." Weather 15, 204. 1960.

5. Changnon, S. A. "Recent Studies of Urban Effects on Precipitation in the United States." Presented at W.M.O. Symposium on Urban Climates and Building Climatology. Brussels, Belgium. Oct. 1968.

6. Commins, B. T., and Waller, R. E. "Observations from a Ten-year Study of Pollution at a Site in the City of London." Atmospheric Environment 1, 49. 1967.

7. Feig, A. M. "An Evaluation of Precipitation Patterns over the Metropolitan St. Louis Area." Proceedings First National Conference on Weather Modification. Albany, N.Y. American Meteorology Society. pp. 210-219.

8. Geiger, R. The Climate near the Ground. Cambridge, Mass. Harvard University Press. 611 p. 1965.

9. Georgii, H. W. "The Effects of Air Pollution on Urban Climates." Presented at W.M.O. Symposium on Urban Climates and Building Climatology, Brussels, Belgium. Oct. 1968.

10. Holzworth, G. C. "Some Effects of Air Pollution on Visibility in and Near Cities." Symposium: Air Over Cities. U.S. Public Health Service, Taft Sanitary Engineering Center, Cincinnati, Ohio, Technical Report A 62-5. pp. 69-88.

11. MIT. Inadvertent Climate Modification. Report of the Study of Man's Impact on Climate (SMIC). The MIT Press. Cambridge, Mass. 1971.

12. Matthews, William H.; Kellogg, W. W.; and Robinson, G. D. (ed.) Man's Impact on the Climate. The MIT Press, Cambridge, Mass. 1971.

13. National Center for Atmospheric Research. Climate Change and the Influence of Man's Activities on the Global Environment. Sept. 1972.

14. Oke, T. R., and Hannell, F. G. "The Form of the Urban Heat Island in Hamilton, Canada." Presented at W.M.O. Symposium on Urban Climates and Building Climatology. Brussels, Belgium. Oct. 1968.

15. Retty, R. M. "Global Climate Changes Possible." ASME Paper 73-Pwr-11.

16. Robinson, E. "Effects on the Physical Properties of the Atmosphere." Air Pollution. ed. by A. C. Stern. v. 1, 2nd ed. New York, Academy Press. 694 p.

General Bibliographical Source

Brooks, C. E. P. "Selected Annotated Bibliography on Urban Climate." Metorology Abstracts Bibliography 3, 734. 1952.

Peterson, James T. The Climate of Cities: A Survey of Recent Literature. National Air Pollution Control Administration, Raleigh, N.C. Oct. 1969. 53 p.

VII. Effects on Resources [4,11,12,13,15,17,22]

A. Degradation of Basic Resources--Reducing the Potential of Water, Air, and Land for:

1. Agricultural use [16]
2. Industrial use
3. Commercial use
4. Municipal use
5. Recreational use
6. Others--consumption, navigation, etc.

B. Depletion of Other Resources

1. Minerals--metals and non-metals
    a. Minerals from the land
    b. Minerals from the sea [16]
2. Chemicals
    a. Inorganic
    b. Organic
3. Construction materials
4. Energy [2,5,6,7,9]
    a. Fossil fuels [14,18]
    b. Coal [1,8,21]
    c. Petroleum
    d. Natural gas [19]
    e. Nuclear [20]
    f. Hydraulic
    g. Solar
    h. Geothermal
    i. Wind

j. Wood or other biological resources

k. Chemical energy resources

l. Tidal resources

5. Forest [10]

C. Key Aspects which Affect the Depletion of Resources

1. Nature of resources

a. Renewable

b. Non-renewable

c. Total amount available

d. Spacial distribution of resources

2. Uptake of resources by human activities

a. Quantity

b. Recycling

3. Technology [3,6]

a. More efficient use of resources

b. Enable recycling of renewable resources

c. Enable the replacement of less abundant resources by the abundant

Effects on Resources

Bibliography

1. Analysis of the Availability of Bituminous Coal in the Appalachian Region. Bureau of Mines, Pittsburg, Pa. July 1971. 77 p.

2. Bobo, D. L.; Keitz, E. L.; Morris, J.; and Yeager, K. E. A Survey of Fuel and Energy Information Sources, Vol. I. Mitre Corp., McLean, Va. Nov. 1970. 298 p.

3. Council on Environmental Quality. Resource Recovery: The State of Technology. U.S. Government Printing Office. Feb. 1973. 67 p.

4. Ehrlich, Paul R., and Ehrlich, Anne H. Population, Resources, Environment--Issues in Human Ecology. W. H. Freeman and Co., San Francisco. 1970.

5. Energy Policy Project. Exploring Energy Choices. March 1974. 88 p.

6. Energy Technology to the Year 2000. A Special Symposium Published by Technology Review. MIT, Cambridge. 1972. 48 p.

7. Hammond, Allen L.; Metz, William D.; and Maugh, Thomas H. III. Energy and the Future. American Association for the Advancement of Science. 1973. 184 p.

8. Hoffman, L.; Lysy, F. J.; Morris, J. P.; and Yeager, K. E. Survey of Coal Availability by Sulfur Content. Mitre Corp., McLean Va. May 1972. 171 p.

9. Kruger, Paul, and Otte, Carel. "The Energy Outlook: Now to 1985." Aware. Sept. 1973. p. 3-7.

10. Manning, Glenn H., and Grinnell, H. Rae. Forest Resources and Utilization in Canada to the Year 2000. p. 42-46. Environment Canada. 1971.

11. Material Needs and the Environment Today and Tomorrow. Final Report of the National Commission on Materials Policy. June 1973.

12. Murdoch, William W. (ed.) Environment, Resources, Pollution, and Society. Sinauer Associates, Inc., Stamford, Connecticut. 1971. 435 p.

13. National Academy of Sciences. Resources and Man. National Research Council, Committee on Resources and Man. W. H. Freeman and Co., San Francisco. 1969.

14. Oil Availability by Sulfur Levels. Bureau of Mines, Washington, D.C. Aug. 1971. 282 p.

15. Ray, G. F. "Energy: Resources and Demand in this Century and Beyond." Long Range Planning. v. 6, #1. Mar. 1973.

16. Research Institute of the Gulf of Maine. Renewable Marine Resources Development Project: Conference Proceedings. Cooperative Extension Service, University of Maine. 1972. 111 p.

17. Ryan, Charles J. Materials, Energy, and the Environment: the Need to Produce, Conserve, and Protect. National Commission on Materials Policy Report. Feb. 1973. 21 p. Special Report.

18. Study of the Future Supply of Low Sulfur Oil for Electrical Utilities. Hittman Associates, Inc., Columbia, Md. Feb. 1972. 76 p.

19. Study of the Future Supply of Natural Gas for Electrical Utilities. Hittman Associates, Inc., Columbia, Md. Feb. 1972. 44 p.

20. Survey of Nuclear Power Supply Prospect. Hittman Associates, Inc., Columbia, Md. Feb. 1972. 99 p.

21. United States Coal Resources and Production. Bureau of Mines, Washington, D.C. June 1971. 51 p.

22. U.S. Environmental Protection Agency. Energy Conservation Strategies. EPA-R5-73-021. July 1973. 114 p.

General References for Environmental Effects

Contra Costa County Planning Department. Land Use Classification Manual. April 1970.

Berkowitz, David A., and Squires, Arthur M. (ed.) Power Generation and Environmental Change. MIT Press, Cambridge, Mass. 1971.

Detwyler, Thomas R. Man's Impact on Environment. New York: McGraw-Hill Book Co. 1971. 709 pp.

Environmental Quality and Social Justice in Urban America. Conference report, sponsored by the Conservation Foundation. Publications Dept., Washington, D.C. 145 pp.

Flack, J. E., and Shipley, M. C. Man and the Quality of His Environment. U. of Colorado Press, Boulder. 1968.

Hodges, Laurent. Environmental Pollution. Holt, Rinehart and Winston, Inc. N.Y. 1973.

Jarrett, H. (ed.) Environmental Quality in a Growing Economy. Baltimore. 1966.

MIT. Man's Impact on the Global Environment--Assessment and Recommendations for Action. Report of the Study of Critical Environmental Problems (SCEP). 1970.

Meier, Richard L. Interpretation: Insights into Pollution. Inst. of Urban and Regional Development, U. of California, Berkeley, Reprint #68. July 1971. 7 pp.

Singer, S. Fred (ed.) Global Effects of Environmental Pollution. Springer Verlag New York Inc., N.Y. 1970.

Thomas, W. L. Jr. (ed.) Man's Role in Changing the Face of the Earth. U. of Chicago Press, Chicago. 1956.

Tuan Yi-Fu. Man and Nature. Commission on College Geography, Resource paper #10. Association of American Geographers, Washington, D.C. 1971. 46 pp.

U.S. Council on Environmental Quality. Energy and the Environment--Electric Power. Aug. 1973. 58 pp.

U.S. Technical Committee on Industrial Classification. Standard Industrial Classification Manual. Office of Statistical Standards, Bureau of the Budget, Executive Office of the President. 1967.

Wilkinson, H. R. Man and the Natural Environment. University of Hull. Occasional Papers in Geography #1. 1966. 32 pp.

VIII. Social and Cultural Effects

A. Population

1. Growth--urban growth, migration
2. Age distribution
3. Sex distribution
4. Ethnic distribution
5. Racial distribution

B. Food and Nutrition

C. Housing

1. Quality
2. Quantity

D. Education

1. Provision for education
2. Equality of opportunity

E. Health and Sanitation

1. Solid waste disposal
2. Water supply
3. Rodent and pest control

F. Security and Safety

1. Traffic hazards
2. Crime
3. Fire hazards
4. Others

G. Mobility

1. Accessibility to work
2. Accessibility to shopping facilities
3. Accessibility to schools

4. Accessibility to business centers
5. Accessibility to cultural centers
6. Accessibility to recreational areas

H. Effects on Cultural Activities

I. Effects on Religious Activities

IX. Effects on Ecosystems*

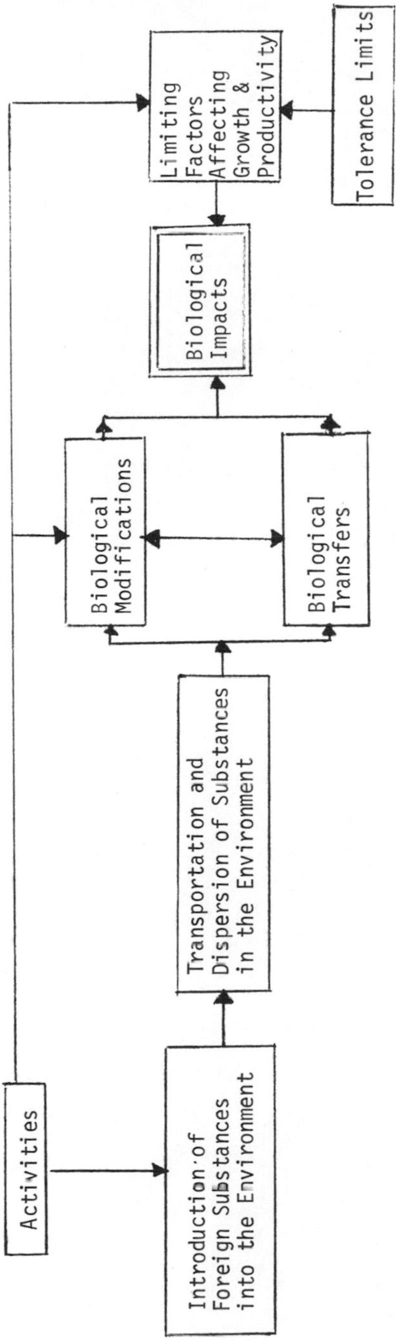

Activities

Introduction of Foreign Substances into the Environment

Transportation and Dispersion of Substances in the Environment

Biological Modifications

Biological Transfers

Biological Impacts

Limiting Factors Affecting Growth & Productivity

Tolerance Limits

*Keyed references in this section are taken from the bibliography of the subject area "Ecology."

A. Types of Ecosystems [69,105]

1. The terrestrial environment

a. Forests [50,78,108]

(1) tropical lowland forests
(2) national forests

b. Watersheds

c. Range and grasslands [49]

(1) savannas
(2) temperate grasslands
(3) semi-arid grasslands

d. Arid and semi-arid regions [17,83]

e. Agricultural

f. High-latitude

2. The freshwater environment [37]

a. Lentic community

b. Lotic community [39]

3. The marine environment [28,65,84,102]

a. Major biomes

b. Sub-biomes

4. The estuarine environment [33,56]

a. Geographic location

b. Type of organisms

B. Activities which Contribute to the Impact of Biological Systems

1. Activities discharging foreign substances into the environment vs. Foreign substances

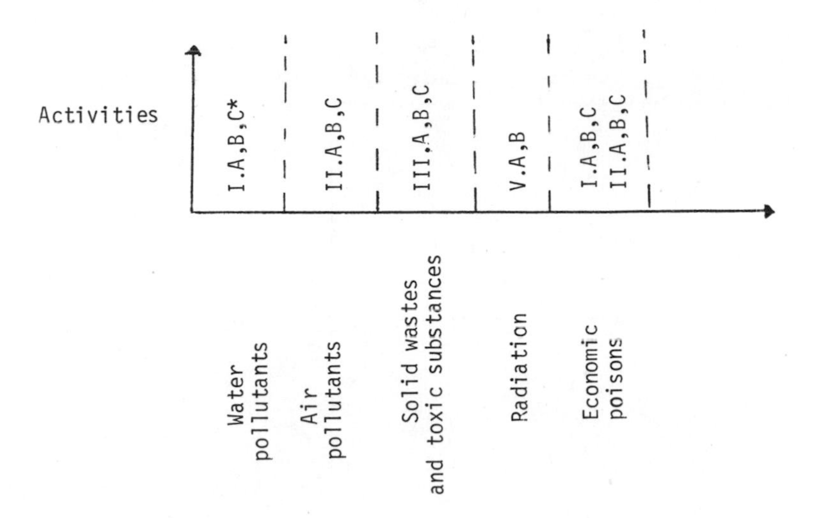

2. Non-polluting activities

   a. Natural phenomena

   b. Recreation

      (1) hunting
      (2) fishing
      (3) snowmobiling

   c. Landscape alteration

      (1) landscaping
      (2) stripmining rehabilitation
      (3) land reclamation
      (4) water resources projects

C. Transfer and Transport of Foreign Substances

   1. Foreign substances in water -- I.D

   2. Foreign substances in air -- II.D

   3. Solid wastes -- III.D

---

*Refers to other items in this subject area

D. Biological Transfers and Modifications*

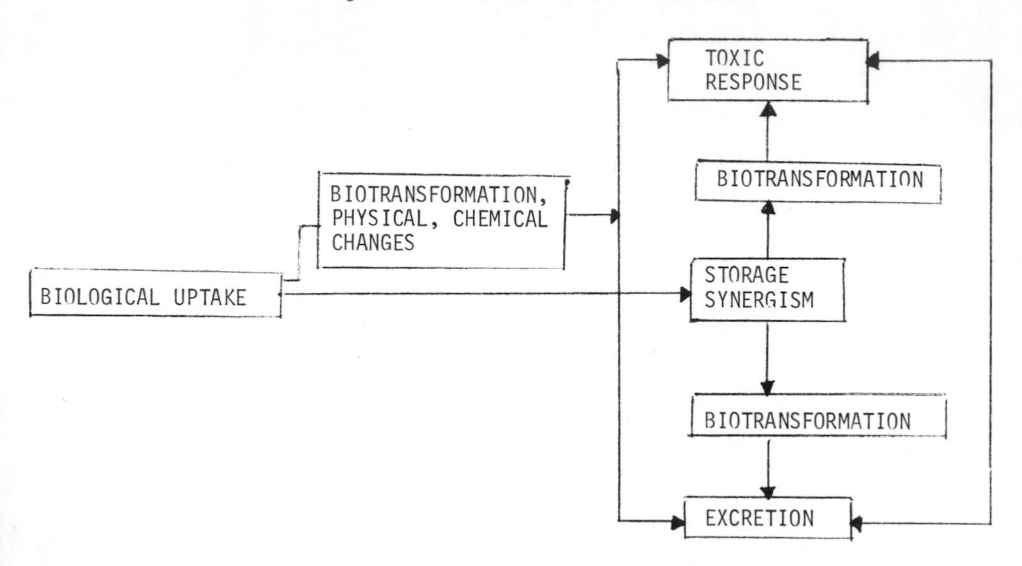

E. Limiting Factors Affecting Growth and Productivity

1. Limiting factors vs. key species vs. tolerance limit

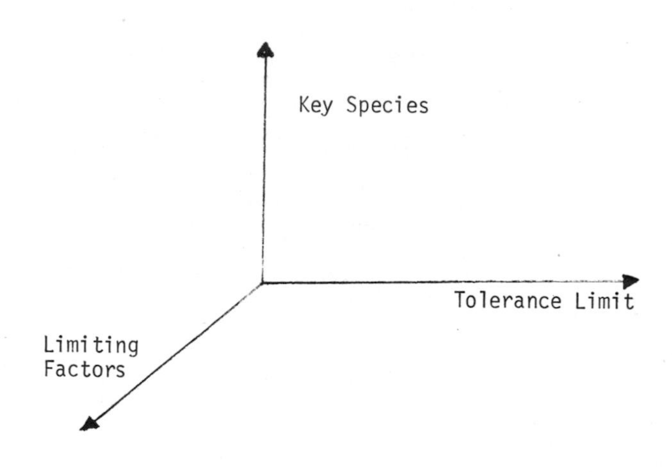

---

*Source: Moore, Stephen F., _et al_. A Preliminary Assessment of the Environmental Vulnerability of Machias Bay, Maine to Oil Supertankers. Dept. of Civil Engineering, MIT, Report No. 162.

2. Limiting factors

a. Abiotic - e.g. marine ecosystem

(1) light

(a) wavelength
(b) intensity
(c) duration

(2) temperature
(3) salinity
(4) pH buffering capacity
(5) noted limiting nutrients

(a) carbon
(b) nitrogen
(c) phosphorous
(d) vitamins
(e) trace metals
(f) sulfur
(g) organic compounds

(6) turbulent mixing
(7) advective transport

b. Biotic

(1) population size
(2) population dynamics

3. Tolerance limits - range within limiting factors produce no biological impact threshold response

F. Biological Impacts[69,105]

1. Impact on energetics[30,48]

a. Biotic energy flows
biomass supported/unit energy flow

b. Food chain

(1) autotrophic organisms
(2) heterotrophic organisms

c. Photosynthesis - respiration
net yield

d. Performance of biological work

e. Biogeochemical cycles

(1) homeostatic mechanisms[87]
(2) organic synthesis
(3) implications of alterations

f. Nutrient cycling [10,36]

(1) mineral cycles and nutrient regeneration
(2) nutrient exchange rate between organisms and the environment

2. Impact on growth and productivity

a. Size of organism

b. Life cycle

c. Niche specialization

d. Growth kinetics

e. Production

f. Metabolic rate

g. Physiology

h. Cell aging rate

i. Behaviour pattern

j. Specialized food

k. Survival

3. Impact on structure [7,38,48,99]

	a. Diversity (1) variety (2) equitability (3) stratification (4) biochemical	b. Resiliency	c. Stability	d. Changes in distribution pattern in space
a. Cell		X		X
b. Organism		X		X
c. Population	X	X	X	X
d. Community	X	X	X	X
e. Ecosystem			X	

4. Impact on the individual in the ecosystem
    a. Ecological niche
        (1) habitat
        (2) trophic level
5. Impact on the organization at the community level
    a. Competition and predator-prey relationships
    b. Ecological succession
6. Impact on the organization at the population level
    a. Population size
        (1) natality
        (2) mortality
        (3) dispersal
    b. Population dynamics [62]
        (1) fluctuation
        (2) competition
        (3) parasitism
    c. Population control
7. Impact on the organization of the ecosystem [87,99]
    a. Homeostasis
        (1) stability to resist external perturbation
        (2) entropy
        (3) internal symbiosis
        (4) succession

ENVIRONMENTAL INDICATORS

As concern of environmental quality increases, there is an awareness growing among planners, decision-makers, and the public that more accurate and objective information on the status and trends of the environment is necessary to improve the formulation, implementation and communication of environmental policy. An environmental indicator, a number derived from a collection of statistics which quantitatively summarizes or measures the condition of the physical, ecological, social, economic or aesthetic environment, may often provide the kind of information required.

This subject area informs an environmental manager of the significant components in each environment where environmental indicators should be or have been formulated to reflect their trends and status. By providing information on who are the potential users of environmental indicators, how they can make use of environmental indicators, and what the limitations are on the use of indicators, the subject area enables the environmental manager to guide himself and others to use indicators discriminately. In some areas where indicators should be but have not been formulated, the subject area outlines the procedures to develop indicators and identifies what is needed in each procedure. Efforts in the development of environmental indicators are included in the last section of the outline. In all, the purpose of the subject area is to acquaint an environmental manager with a set of potentially effective tools for monitoring the status and trends of the environment and for evaluating the effectiveness of environmental policy.

The subject area is related to other subject areas and in particular Modeling and Monitoring. Relevant sections of other subject areas should be reviewed simultaneously in order to better understand and to make full use of this area.

ENVIRONMENTAL INDICATORS

General Outline

Needs of and Procedures for Developing Indicators

Types of Indicators

The Users, Uses, and Limitations of Indicators

Indicators Currently Used and Proposed

ENVIRONMENTAL INDICATORS

Detailed Outline

I. Definition - a number derived from a collection of statistics which quantitatively summarizes or measures the condition of the physical, ecological, social, economic or aesthetic environment.

II. Types of Indicators [20,26,27,44,46]

A. Indicators for the Physical Environment [17,23,29,30,31,34,39,42,49,50]

1. Pollution [3,4,34]

a. Water quality [8,13,18]

b. Air quality [19,25,30,45]

c. Characteristics of land [23]

d. Weather and climate

2. Depletion of resources

a. Media as a resource

(1) air
(2) water
(3) land

b. Others

(1) food
(2) minerals
(3) chemicals
(4) construction materials
(5) energy

B. Indicators for the Ecological Environment

1. Diversity

a. Variety

b. Equitability

c. Stratification

d. Biochemical

*Sections of this subject area were extracted from the draft outline of the Environmental Quality Indicators Planning Study Report. Environmental Studies Board, National Academy of Sciences. Washington, D.C. 20418

2. Resiliency
3. Stability
4. Changes in distribution pattern in space

C. Indicators for the Economic Environment

1. Total activity and income
2. Markets and production
3. Currency
4. Employment rate
5. Financial debts and surpluses

D. Indicators for the Social Environment [1,2,5,6,7,9,14,15,22,24,40,52]

1. Individual dignity
2. Equality
3. Conflicts and tensions
4. Safety
5. Health
6. Housing
7. Employment
8. Utilities
9. Transportation
10. Welfare
11. Education
12. Amenity
13. Recreation [30,37]
14. Cultural activity

E. Indicators for the Aesthetic Environment

1. Scale
2. Color

3. Texture

4. Pattern

5. Variety

6. Tranquility

III. The Uses of Indicators

A. Users and Uses [28]

Users	Uses
1. Public	1. To keep informed of the status and trend of the environment 2. To provide the basis for public intervention 3. To allow the public to participate more effectively in a. principle setting b. formulation of goals c. formulation of objectives d. evaluation of alternatives 4. To evaluate the responsiveness of public officials 5. To attribute the causes of environmental consequences
2. Decision makers policy makers administrators program planners	1. To assess environmental problems 2. To formulate goals and objectives 3. To develop policy approaches and define alternative programs 4. To provide the basis for evaluating alternatives

	5. To aid in program implementation a. setting stages for implementation b. setting priorities among program components c. allocating resources for program components 6. To aid in program evaluation 7. To indicate trends that may require new control
3. Judicial system	1. To establish standards 2. To recognize existence and extent of violations 3. To provide objective information to settle disputes
4. Non-governmental and business groups	
5. Conservation groups	Same as Public
6. Researchers	To alert the public and the decision makers of the status and the trend of the environment

B. Limitations on the Use of Indicators

1. No guarantee that indicators will be heeded by potential users
2. Obscuring the importance of substantial subjective components used in formulating indicators
3. Misuses of indicators by users due to
    a. Misunderstanding the meaning of the indicators used
    b. Overlooking the assumptions and necessary conditions upon which the indicators are formulated

IV. Development of Indicators - Procedures and Needs [1,16,27,29,35,38]

Procedures	Needs
1. Basic data gathering	1. Measuring methods 2. Monitoring systems a. types of monitoring systems (Monitoring II*) b. technical approaches in monitoring systems (Monitoring III) c. appropriateness criteria for the choice of monitoring systems (Monitoring IV) d. operational criteria for the choice of monitoring systems (Monitoring V) e. Monitoring systems vs. Appropriateness and Operational criteria (Monitoring VI) f. criteria for cost-effective monitoring programs for developing countries (Monitoring VII) g. systems and trends that can be monitored (Monitoring IX)
2. Data aggregation	Model development 1. Types of modelling (Modelling ID) 2. Models available for use in analyzing environmental systems (Modelling II) 3. Appropriateness criteria for model choice (Modelling III) 4. Operational criteria for model choice (Modelling IV) 5. Models available for use vs. Appropriateness and Operational criteria (Modelling V) 6. Candidates for modelling (Modelling VI)

*Refers to items in the subject area on "Monitoring."

3. Indicators Formulation	1. Extensive experimentation and feedback from models developed 2. To identify critical factors
4. Dissemination of Indicators	1. Collation and documentation of environmental statistics and indicators 2. Means to classify indicators according to the requirements and sophistication of users

V. Efforts in U.S.A. to Develop Environmental Indicators

A. Federal Governmental Efforts

1. Congressional efforts

   a. House Resolution (HR) - 56 [48]

   b. Library of Congress survey, Dec. 1973 [30]

2. Federal agencies

   a. Council on Environmental Quality (CEQ) annual reports [10,11]

   b. Environmental Protection Agency (EPA) - Quality of Life Indicators [7]

   c. Northeast Regional Office of the Bureau of Outdoor Recreation, U.S. Department of the Interior (DOI)

3. State and regional efforts

   a. North Carolina - Planning for Environmental Quality [35]

   b. San Diego - Integrated Regional Environmental Management Program (IREM) [38]

4. International - Environment Canada [20]

5. Other efforts

   a. American Association for the Advancement of Science (AAAS) Symposium, 1971 [44]

   b. National Wildlife Federation - Environmental Quality (EQ) Index [31,32]

c. MITRE Corp. - "Monitoring the Environment of the Nation" [29]

d. Midwest Research Institute - "Quality of Life in the U.S., 1970"[24]

B. Environmental Categories

1. Air quality

a. MITRE Air Quality Index (MAQI) [25]

b. Extreme Value Index (EVI) prepared by MITRE for CEQ [29]

c. Oak Ridge Air Quality Index (ORAQI) - Oak Ridge National Laboratory [45]

2. Water quality

a. Prevalence-Duration-Intensity (PDI) - EPA [8,18,51]

b. Enviro Control Water Pollution Analysis - CEQ [13]

c. Water Quality Index (WQI) - National Sanitation Foundation [8]

3. Pesticides

"Environmental Indicators of Pesticides" - Stanford Research Institute [42]

4. Land use [36,37]

a. Supply

b. Needs

c. Demand

d. Economic

5. Radiation

Cumulative Exposure Index (CUEX) - Oak Ridge National Lab [39]

6. Oceans

National Academy of Sciences (NAS) 1973 study on Predicting Ocean Pollutants

7. Noise

a. Lipscomb - AAAS Symposium [44]

b. "Community Noise Scale" - EPA [49,50]

c. Noise Index (NI) - MITRE [17]

8. Wildlife

   Endangered species list - DOI

9. Recreation

   a. Recreational Availability Indicators - MITRE [4,29]

   b. Urban Area Recreation Indices - MITRE [4,29]

10. Environmental perception

    a. "Environmental Psychology" - Kenneth Craik [12]

    b. Bart Rapid Transit System Impact Study - University of California, Berkeley

C. Social Indicators

1. "Social Indicators 1973" - U.S. Office of Management and Budget (OMB) [41]

2. "Towards a Social Report" - U.S. Department of Health, Education, and Welfare (HEW) [47]

3. New York Times Article [33]

4. Center for Coordination of Research on Social Indicators - Social Science Research Council (SSRC)

5. Urban environment

   a. "Quality of Life in Metropolitan Washington, D.C." - Urban Institute [21]

   b. "Systematic Measurement of the Quality of Urban Life" - Los Angeles Community Analysis Bureau (CAB) [43]

   c. "Quality of Life in Urban America - New York City - A Regional and National Comparative Analysis"[2]

ENVIRONMENTAL INDICATORS

Bibliography

1. Bauer, Raymond A. (ed.) Social Indicators. The M.I.T. Press, Cambridge, Massachusetts. 1966.

2. Berenyi, J. (ed.) The Quality of Life in Urban America. New York City: A Regional and National Comparative Analysis. Vol. 1. Office of the Mayor, New York. May 1971.

3. Bisselle, Charles, et al. Environmental Trends = Radiation, Air Pollution, Oil Spills. MTR-6013. The Mitre Corporation, McLean, Virginia. May 1971.

4. Bisselle, Charles A., et al. National Environmental Indices: Air Quality and Outdoor Recreation. MTR-6159. Mitre Corporation, McLean, Virginia. April 1972. 263 pp.

5. Bonjean, Charles; Hill, R.; and McLemore, S. Sociological Measurement. An Inventory of Scales and Indices. Chandler Publishing Co., San Francisco. 1967. 580 p.

6. Brady, Henry. Social Indicators. Divison of Social Sciencies, National Science Foundation. 1970.

7. Brossman, Martin W. Quality of Life Indicators. A Review of State-of-the-Art and Guidelines Derwed to Assist in Developing Environmental Indicators. U.S. Environmental Protection Agency, Environmental Studies Division. Dec. 1972. 88 p.

8. Brown, R. "A Water Quality Index--Crashing the Psychological Barrier." Indicators of Env. Quality. Ed. by William A. Thomas. Plenum Press, New York. 1972. p. 173.

9. Cassidy, Michael W. A. Social Indicators: Accidents and the Home Environment. Institute of Urban and Regional Development, University of California, Berkeley. Working Paper No. 132. Oct. 1970. 30 pp.

10. Council on Environmental Quality. Environmental Quality. The Third Annual Report of the Council on Environmental Quality. U.S. Government Printing Office, Washington, D.C. 1972.

11. Council on Environmental Quality. Environmental Quality. The Fourth Annual Report of the Council on Environmental Quality. U.S. Government Printing Office, Washington, D.C. 1973.

12. Craik, Kenneth H. "Environmental Psychology." Annual Review of Psychology, 24: 403-421. 1973.

13. Enviro Control Inc. National Assessment of Trends in Water Quality. 1972.

14. Flax, Michael J. A Study in Comparative Urban Indicators: Conditions in 18 Large Metropolitan Areas. The Urban Institute, Washington, D.C. April 1971.

15. Flax, Michael J. Future Prospects for the Development of Additional Social and Urban Indicators. The Urban Institute, Washington, D.C. July 1971.

16. Garn, Harvey A. and Flax, Michael J. Urban Institute Indicator Program. The Urban Institute, Washington, D.C. July 1971.

17. Goldstein, S. Environmental Noise Quality. MITRE Corporation. Report M71-8. 1971.

18. Greeley, R. S., et al. Water Quality Indices. M72-54. The Mitre Corporation, McLean, Virginia. April 1972.

19. Heck, W. W. "The Use of Plants as Indicators of Air Pollution." Air and Water Pollution, 10. February 1966. pp. 99-111.

20. Inhaker, Herbert. A National Environmental Quality Index for Canada--Technical Edition. Planning and Finance Service, Environment Canada, Ottawa, Ontario.

21. Jones, Martin V. and Flax, Michael J. The Quality of Life in Metropolitan Washington, D.C.--Some Statistical Benchmarks. The Urban Institute, Washington, D.C. March 1970.

22. Kriezer, Martin H. Social Indicators for the Quality of Individual Life. Institute of Urban and Regional Development, University of California, Berkeley. Working Paper No. 104. Oct. 1969. 28 pp.

23. Land Use Indicators of Environmental Quality. An Examination of Existing Federal Data and Future Needs. Earth Satellite Corp., Washington, D.C. April 15, 1972. 152 p.

24. Liw, Ben-Chieh. The Quality of Life in the United States. 1970 Index, Rating and Statistics. Midwest Research Institute, Kansas City, Missouri. July 1973.

25. Lubore, S. H. and Pikell, R. P. Indices of Air Quality. M72-63. The Mitre Corporation, McLean, Virginia. May 1972.

26. Lyndon Baines Johnson School of Public Affairs Community Analysis Research Project. An Introductory Set of Community Indicators. Lyndon Baines Johnson School of Public Affairs, University of Texas, Austin. 1973. 46 pp.

27. Lyndon Baines Johnson School of Public Affairs Community Analysis Research Project. A Resource Handbook for Developing Community Indicators. Lyndon Baines Johnson School of Public Affairs, University of Texas, Austin. 1973. 52 pp.

28. MacDonald. "Uses of Environmental Indices in Policy Formulation." Indicators of Environmental Quality. Ed. by W. Thomas. Plenum Press, New York. 1972.

29. Mitre Corporation. Monitoring the Environment of the Nation. MITRE Corporation. Report MTR-1600. April 1971.

30. National Environmental Policy Act of 1969. Environmental Indices--Status of Development Pursuant to sections 102(2)(B) and 204 of the Act. Environmental Policy Division, Congressional Service, Library of Congress. Dec. 1973.

31. National Wildlife Federation. 1971 EQ Index Reference Guide. National Wildlife Federation, Washington, D.C. 1971.

32. National Wildlife Federation. "National Environmental Quality Index." National Wildlife Magazine. October-November 1971.

33. NY Times. A recent NY Times article surveys developments in the field and includes commentary concerning the goals and the methods of prominent researchers. For more information contact Laurie Goldstein and David Holtz, Staff of the Environmental Studies Board, National Academy of Sciences, Washington, D.C. 20418.

34. Padgett, J. H. and Stanford, R. A. "An Industrial Pollution Index." Water Resources Bulletin. April 1973. v. 9. n. 2. p. 320.

35. Paul, Roy A. Planning for Environmental Quality, Phase I and II. State Planning Division, Dept. of Administration, North Carolina. June 1972.

36. Pikul, Robert, et al. "Development of Environmental Indicators." Indicators of Environmental Quality. Ed. by William Thomas. Plenum Press, New York. 1972.

37. Pikul, Robert, et al. Development of Environmental Indices: Outdoor Recreational Resources and Land Use Shift. Report #M71-72. The Mitre Corporation, McLean, Virginia. Dec. 1971.

38. Research Analysis Corporation. Environmental Quality Index--A Feasibility Study. Environmental Development Agency, County of San Diego, San Diego, California. June 1972. IREM Report.

39. Rohwer and Stherness. "Environmental Indices for Radioactivity Releases." Indicators of Environmental Quality. Ed. by William Thomas. Plenum Press, New York. 1972. p. 249.

40. Sheldon, Eleanor and Moore, Wilbert E. (Eds.) Indicators of Social Change, Concepts and Measurements. The Russell Sage Foundation, New York. 1968.

41. Social Indicators, 1973. U.S. Office of Management and Budget. 1973.

42. Strickland, John and Blue, Thomas. Environmental Indicators for Pesticides. Stanford Research Institute, Menlo Park, California. April 1972. 129 pp.

43. "Systematic Measurement of the Quality of Urban Life--Prerequisite to Management." Los Angeles Community Analysis Bureau (CAB).

44. Thomas, William A. (ed.) Indicators of Environmental Quality. Plenum Press, New York. 1972.

45. Thomas, W., et al. Air Quality Index. Oak Ridge National Laboratory, Report #ORNL-NSF-EP-8. 1971.

46. Thomas, William A., et al. Biological Indicators of Environmental Quality. A Bibliography of Abstracts. Ann Arbor Science Publishers, Inc., Ann Arbor, Michigan. 1972.

47. Towards a Social Report. Department of Health, Education, and Welfare, Washington, D.C. 1969.

48. U.S. Congress. Environmental Data Systems Act (HR-56).

49. U.S. Environmental Protection Agency. Community Noise. Office of Noise Abatement and Control. Technical Document NTID 300.3. Dec. 31, 1971.

50. U.S. Environmental Protection Agency. Fundamentals of Noise--Measurement, Rating Schemes, and Standards. Office of Noise Abatement and Control. Technical Document NTID 300.15. Dec. 31, 1971.

51. U.S. Environmental Protection Agency. The Economics of Clean Water. Vol. I & II. 1972.

52. Wilcox, Lester D.; Brooks, Ralph M.; Beal, George M.; and Klonglan, George E. Social Indicators and Societal Monitoring. An annotated Bibliography. Jossey-Bass, Inc., San Francisco. 1972.

ENVIRONMENTAL IMPACT ASSESSMENT METHODOLOGY

For decades the quantifiable costs and benefits, generally expressed in terms of dollars, of an activity have dominated the evaluation of the activity--particularly those that were undertaken to satisfy material wants. Values which could not be quantified or expressed in terms of dollars were only given superficial treatment and very little weight in decision-making. Environmental quality is one such value.

Many developed countries have had tremendous economic development but not without deterioration of their environment. The importance of values of environmental quality began to grow when the degradation of the environment became increasingly tangible. The awareness of environmental quality gradually pushed planners and decision-makers alike to change their concept of activity evaluation. Now emerging is a new concept that requires all probable impacts of an activity, quantifiable or unquantifiable, direct or indirect, to be evaluated in decision-making.

To make the new concept operative, an environmental manager must be able to systematically identify, predict, and evaluate all probable impacts of an activity. This subject area outlines the procedures to analyze objectively the impacts of an activity and to evaluate alternatives. These procedures can also help the environmental manager to be aware of what information he needs to select alternatives. Making an objective analysis of all probable impacts and evaluating alternatives in this sense are no easy tasks. This subject area provides the environmental manager insight on what obstacles may limit either process.

Although techniques to overcome these obstacles may not be available at present, the environmental manager is at least aware of where the gaps are and can focus his efforts on these gaps.

This subject area is closely related to the subject area of Environmental Effects which broadens the environmental manager's perspectives on the spectrum of effects which an activity may have on the environment, and the subject area of Environmental Indicators which acquaints him with a set of potentially effective tools. Together they augment the environmental manager's ability to assess the environmental impacts of an activity and to select the best choice.

ENVIRONMENTAL IMPACT ASSESSMENTS

General Outline

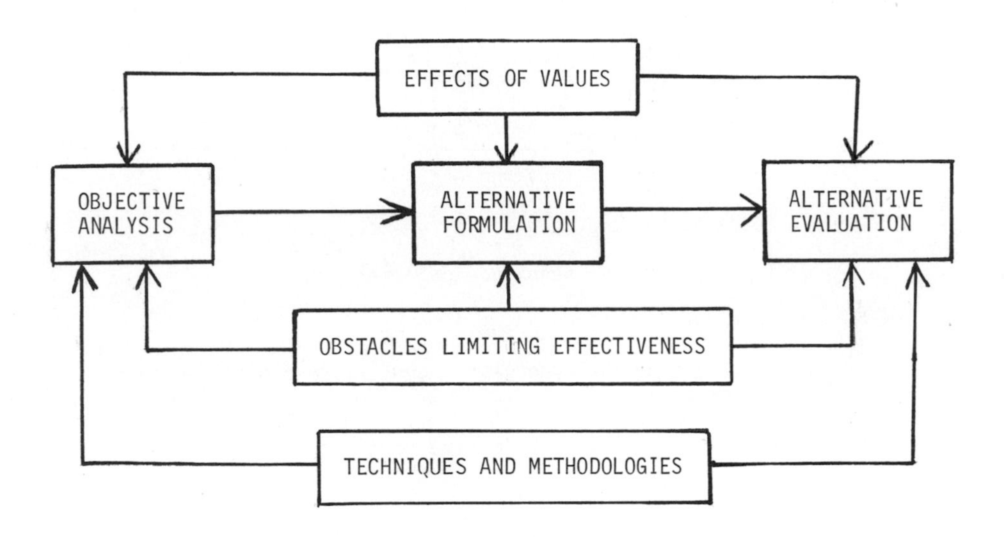

ENVIRONMENTAL IMPACT ASSESSMENT METHODOLOGY

Detailed Outline

I. Objective Analysis *[6,9,12,16,21,22,31,47,48,49,50]

A. Procedure

1. To identify probable impacts
2. To determine what data to gather
3. To determine what environmental indicators to use
4. To determine the probability of occurrence of impacts
5. To determine the magnitude of impacts
6. To determine the dynamics of impacts
7. To determine the synergistic and antagonergistic effects of impacts

B. The Effect of Values on Objective Analysis [23,34]

1. Values sometimes dictate what impacts to be investigated
2. Professional values affect the types of techniques to be used in objective analysis
3. Objective analysis incorporates assumptions based on professional experience and values

C. Obstacles Limiting Objective Analysis [20]

1. Cost of analysis
2. Time constraint
3. Limit of science and technology
4. Problems with data
   a. Inadequate data
   b. Difficulties in obtaining data
      (1) lack of centralization of information
      (2) proprietary information
   c. Conflicting data--uncertainty with information base
5. Problems with modelling and monitoring

---

*The "effects of values" are treated in the subject area on "Values and Perception."

II. Alternatives Evaluation [10,11,29,30,43,44,46]

A. Procedure

1. To characterize impacts with respect to principles, goals, and objectives

2. To determine the weight of various impacts with respect to the degree of concern

3. To evaluate the overall merits and demerits of alternatives (for alternatives evaluation)

B. The Effects of Values on Alternatives Evaluation

1. Impacts may be characterized differently according to values

2. Weights of various impacts assigned differently according to values

3. Professional values affect the types of techniques used to assess alternatives or to define environmental problems

C. Obstacles Limiting Alternatives Evaluation [34,36]

1. Problems in disseminating the results of assessment

   a. Format difficulty

   b. Results incomprehensible to the public and decision makers

2. Problems with the assessors

   a. Lack of disciplinary diversity

   b. Barrier among disciplines

      (1) academic
      (2) jargon

   c. Personal bias

3. The difficulty to put weights on various impacts when degree of concern changes over time; the difficulty to put different weights on a single impact which varies with time

4. The difficulty to incorporate the probability of occurrence of impacts into evaluation

5. The difficulty in putting weight on impacts because of different degree of concern among individuals

III. Environmental Assessment Methodologies [13]

A. Coastal Zone Study [38,39,40]

B. Atomic Energy Commission [45]

C. U.S. Geological Survey [24,25,26]

D. Bechtel Environmental Assessment Matrix [54]

E. Battelle Environmental Evaluation Systems [14,15,53]

F. Optimum Pathway Matrix Technique [52]

G. STV Environmental Assessment [41]

H. Stanford Research Institute--Aesthetic Assessment Methodology [4,5]

ENVIRONMENTAL IMPACT ASSESSMENT METHODOLOGY

Bibliography

1. Appleyard, Donald and Carp, Frances M. The Bart Residential Impact Study: A Longitudinal Empirical Study of Environmental Impact. Institute of Urban and Regional Development, University of Calif., Berkeley. Working Paper No. 205/Bart 12. Feb. 1973.

2. Appleyard, Donald and Okamoto, Rai Y. Environmental Criteria for Ideal Transportation Systems. Institute of Urban and Regional Development, University of Calif., Berkeley. April 1968.

3. A Guide to Social and Environmental Consideration in Transportation Decision-Making (Draft). Urban Systems Laboratory, MIT, Cambridge, Massachusetts.

4. Bagley, M. D. Aesthetic Assessment Methodology--A Case Study. Technical Note TN-OED-011, Stanford Research Institute, Menlo Park, Calif. Nov. 1972.

5. Bagley, M. D. Aesthetic Assessment Methodology for Environmental Impact Analysis. Technical Note TN-OED-004, Stanford Research Institute, Menlo Park, Calif. March 1972.

6. Baumgold, Marian S. and Enk, Gordon A. (ed.) Toward a Systematic Approach to Environmental Impact Review. The Institute on Man and Science, Rensselaerville, N.Y. June 1972. 59 p.

7. Best, Judith A. NEPA Impact Statements. Agency Efforts to Escape the Burden. Cornell University, Ithaca, N.Y. June 1972. 21 p.

8. Best, Judith A. The National Environmental Policy Act as a Full Disclosure Law. Cornell University, Ithaca, N.Y. Dec. 1972. 27 p.

9. Bishop, A. Bruce. An Approach to Evaluating Environmental, Social and Economic Factors in Water Resources Planning. Water Resources Bulletin, American Water Resources Association. vol. 8. no. 4. Aug. 1972. p. 724-735.

10. Bishop, A. Bruce. Public Participation in Environmental Impact Assessment. New England College (presented at the Engineering Foundation Conference on Preparation of Environmental Impact Statements) 1973. 21 pp.

11. Coomber, Nicholas H. and Biswas, Asit K. Evaluation of Environmental Intangibles. Genera Press, Bronxville, N.Y., U.S.A.

12. Cross, Frank L. Jr. "Assessing Environmental Impact." Pollution Engineering. June 1973. v. 5. n. 6. p. 34.

13. Daetz, D. and Schlesinger, B. "A Conceptual Framework for Applying Environmental Assessment Matrix Techniques." Journal of Environmental Sciences. Vol. XVI. no. 4. July/August 1973. p. 11-16.

14. Dee, Norbert and Drobny, Neil L. Environmental Assessments for Effective Water Quality Management Planning. Battelle Columbus Laboratories, Columbus, Ohio U.S.A. April 1972.

15. Dee, Norbert, et al. Environmental Evaluation System for Water Resource Planning. Battelle Columbus Lab., Columbus, Ohio. 1972. 188 p.

16. Development of a Methodology to Classify Beach Resources. Florida Division of State Planning. 1966.

17. Dickert, Thomas G. and Domeny, Katherine R. Environmental Impact Assessment: Guidelines and Commentary. University Extension, University of Calif., Berkeley. 1974.

18. Ditton, Robert B. and Goodale, Thomas I. Environmental Impact Analysis: Philosophy and Methods. The University of Wisconsin Sea Grant Program, National Sea Grant Program, U.S. Dept. of Commerce.

19. Fitch, J. M. "Experimental Bases for Aesthetic Decision." Environmental Psychology: Man and His Physical Setting. H. M. Prochavsky, et al. ed. Holt, Rinehart, Winston, New York. 1970.

20. How Effective are Environmental Impact Statements? National Technical Information Service Report No. P6-230-702/3WP. 85 p. (U.S. Government).

21. Kerri, Kenneth D. "Environmental Assessment of Resource Development" Proc. ASCE. Journal of Sanitary Engineering Div. SA2. April 1972. p. 361-374.

22. Kneese, Allen V. and Bower, Blair T. (ed.) Environmental Quality Analysis--Theory and Method in the Social Sciences. The Johns Hopkins Press, Baltimore, Maryland U.S.A.

23. Lee, H. Perception and Aesthetic Value. Prentice-Hall, New York. 1938.

24. Leopold, L. B. "Landscape Aesthetics, How to Quantify the Scenics of a River Valley." Natural History. vol. 78. Oct. 1969. pp. 36-45.

25. Leopold, L. B. Quantitative Comparison of Some Aesthetic Factors Among Rivers. Geological Survey Circular 620, U.S. Dept. of the Interior.

26. Leopold, L. B., et al. A Procedure for Evaluating Environmental Impact. Geological Survey Circular 645. U.S. Dept. of the Interior.

27. Litton, R. B. "An Aesthetic Overview of the Role of Water in the Landscape." Prepared for the National Water Commission in association with the Dept. of Landscape Architecture, University of Calif., Berkeley. July 1971.

28. Matuszeski, William and Jenny, Brian. "Looking at the Impact Statements." Pollution Engineering. v. 5, n. 5. May 1973. p. 23.

29. Miller, J. R. III. A Systematic Procedure for Assessing the Worth of Complex Alternatives. Mitre Corp., Bedford, Mass. 1967.

30. Odum, Eugene P., et al. Totality Indices for Evaluating Environmental Impact: A Test Case- Relative Impact of Highway Alternatives. Use of a Simple Linear Analysis Provides a Means for an Objective Quantitation of Environmental Impact. University of Georgia, Athens, Institute of Ecology, Georgia State Dept. of Transportation. March 1973. 24 p.

31. Ortolano, L. "Issues in Water Resources Impact Assessment." Proc. ASCE. Journal of the Hydraulics Div. Vol. 100. NaHY1. Jan. 1974. p. 173-187.

32. Pill, J. "The Delphi Method: Substance, Context, a Critique, and an Annotated Bibliography." Socio-econ. Plan. Sc. 5,57-71. 1971.

33. Prall, D. W. Aesthetic Judgement. Thomas Y. Crowell, New York. 1929.

34. Quade, R. S. On the Limitation of Quantitative Analysis. Rand Corp. P.4530, Santa Monica, Calif. 1970.

35. Sanford, Fidell; Jones, Glenn; and Pearsons, Karl S. Feasibility of a Novel Technique for Assessing Noise--Induced Annoyance. Bolt, Beranek and Newman, Inc. Canoga Park, Calif. Sept. 1973. 107 p.

36. Smallwood, Jessie Marie. Problems in Preparing an Environmental Impact Statement: the Social, Displacement and Relocation Section. Arthur D. Little, Inc. 8 p.

37. Society and the Assessment of Technology. OECD Publications Center, Suite 1207, 1750 Pennsylvania Ave. N.W. Washington, D.C. 20006.

38. Sorenson, Jens C. A Framework for Identification and Control of Resource Degradation and Conflict in the Multiple Use of the Coastal Zone. Dept. of Landscape Architecture, College of Env. Design, University of Calif., Berkeley. Dec. 1971. 40 p.

39. Sorenson, Jens and Pepper, James. Procedure for Regional Clearing-House Review of Environmental Impact Statements. Prepared for the Association of Bay Area Governments. April 1973. 225 p.

40. Sorensen, Jens C. and Moss, Mitchell L. Procedures and Programs to Assist in the Environmental Impact Statement Process. USC-SG-AS2-73. April 1973. 38 p.

41. Stover, Lloyd V. Environmental Impact Assessment: A Procedure. STV. Inc., Pottstown, Penn. May 1972. 25 p.

42. The River Basin Model: Assessment Department. Environmetrics. Inc. Washington, D.C. Dec. 1971. 89 p.

43. U.S. Army Engineering Institute of Water Resources. Analyzing the Environmental Impacts of Water Projects. Ed. by L. Ortolano. Report #73-3, Ft. Belvoir, Virginia. 1973.

44. U.S. Army Engineering Institute of Water Resources. An Analysis of Environmental Statements for Corps of Engineers. Water Projects' Report #72-3, Ft. Belvoir, Virginia. 1972.

45. U.S. Atomic Energy Commission. Directorate of Regulatory Standards. Regulatory Guide 4.2: Preparation of Environmental Reports for Nuclear Power Plants. March 1973.

46. U.S. Council on Environmental Quality. "Preparation of Environmental Impact Statements: Guidelines." Federal Register. Vol. 38. No. 147. p 20549-20562. August 1, 1973.

47. U.S. Dept. of the Army. Planning, Preparation and Coordination of Environmental Statements. ER 1105-2-507. Washington, D.C. Feb. 1973.

48. U.S. Environmental Protection Agency, Region 1. Environmental Planning and Assessments for Water Quality Management Plans and Projects. Feb. 1973. 29 p.

49. U.S. Environmental Protection Agency, Region X. Environmental Impact Statement Guidelines. April 1973. 123 p.

50. U.S. Water Resources Council. "Establishment of Principles and Standards for Planning Water and Related Land Resources." Federal Register. Vol. 38. No. 174. Sept. 10, 1973. p. 24777-24869.

51. University of Calif. Coastal Zone Bibliography: Citations to Documents on Planning, Resources Management and Impact Assessment. Institute of Marine Resources, University of Calif. Seagrant Publication No. 8. Aug. 1973. 89 p.

52. University of Georgia. Optimum Pathway Matrix Analysis Approach to the Environmental Decision Making Process. Institute of Ecology, Athens, Georgia. 1971.

53. Whitman, I., et al. Design of an Environmental Evaluation System. Battelle Columbus Lab., Columbus, Ohio. 1971. 61 p.

54. Schlesinger, B., and Hughes, R. A. "Environmental Assessment of Alternative Shipbuilding Sites." Bechtel Corporation, San Francisco, Calif. Oct. 1972.

MODELING

The process of modeling environmental systems in some form has always been an important aid to environmental management. In the past, mental models were often cited directly under the claim of professional experience in order to justify certain particular assessments or policies. Recently, the technical state-of-the-art has developed rapidly and now includes extensive and complex dynamic programming techniques applied to simulate various environmental parameters on a large scale. The concept of modeling capability as an important tool that assists decision-making is thus quite valid. However, considerable difficulties can arise regarding how one gets specifically involved in the choice of models for a given problem. The well-informed environmental manager should have before him a comprehensive and previously studied list of criteria relating to the general appropriateness of each existing type of modeling system. He should also be well aware of a number of operational criteria that relate to model performance in general. The approach given in the outline for this subject area is a first step towards providing a grasp of these issues in a straightforward manner.

Given the recognized usefulness of modeling, it should be understood that some modeling techniques can be flexible enough to be used at almost any level of action. Modeling thus ties in with other subject areas to be considered in professional environmental management. For example, the tandem use of modeling for analysis and monitoring for verification of various technical parameters is one important way to determine the effectiveness of a given administrative control procedure.

The role of modeling is that of a technical tool to be used in conjunction with staff assistants and advisors to the decision-maker. Only when those directly involved with decision-making feel comfortable with these techniques will they really be used efficiently and purposefully. Thus, the approach of the outline for this subject area focuses on direct participation of the manager in the choice and use of various environmental models.

MODELING

General Outline

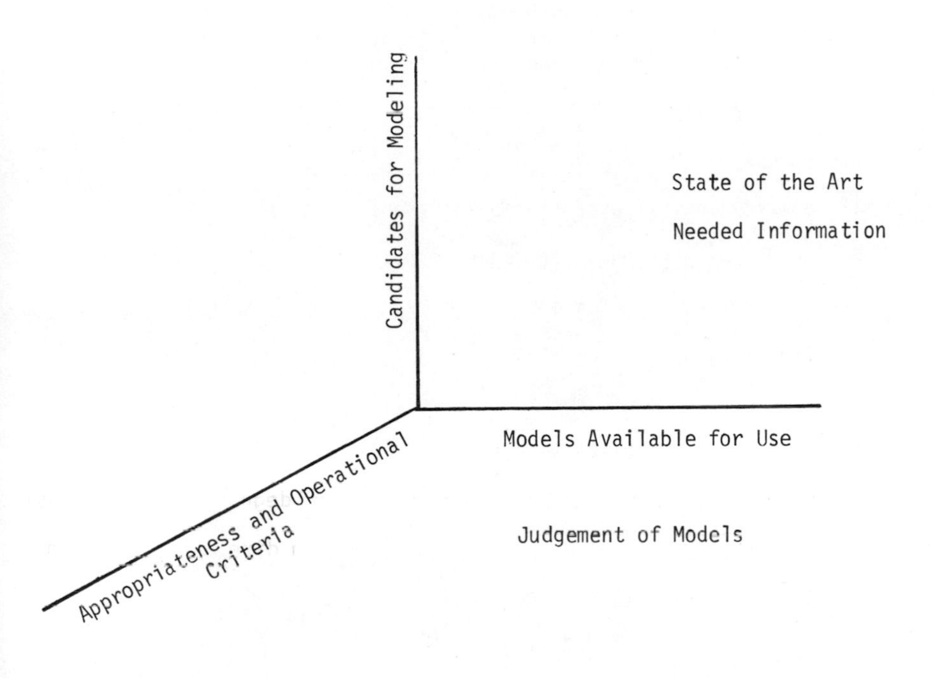

MODELING

Detailed Outline

I. Nature of Modeling

A. Definitions of Modeling

1. Model: An abstract, formal representation of theory about, or empirical observation of some system [21]

2. Other definitions

B. Advantages for Using Models [21]

1. Assists logical organization of ideas

2. Leads to improved system understanding

3. Illustrates need for detail and relevance

4. Expedites speed of analysis

5. Provides framework for testing modifications

6. Easier to manipulate than system itself

7. Permits varying degree of control over system behavior

8. Less costly than direct work with system

C. Historical Development of Modeling

D. General Categories of Modeling [12,17,21]

1. General techniques

a. Physical

b. Schematic

c. Symbolic

2. Different approaches to detail

a. Analytical

b. Numerical

3. Relationship among variables

a. Deterministic

b. Stochastic

c. Mixed

4. Substantive scope of modeling

a. Disciplinary

b. Interdisciplinary

E. Features of a System to be Modeled

1. Boundaries of system

2. Subsystems within systems

II. <u>Models Available for Use in Analyzing Environmental Systems</u>

A. Traditional Computations

B. Analytic Mathematic Optimization Models [15,16]

1. Linear programming

2. Dynamic programming

3. Other non-linear programming

C. Physical System Simulation Models (Using Laboratory Computers) [21]

1. Use of discrete event simulation

2. Programming languages

3. Stochastic concepts

a. Random numbers

b. Other

D. Econometrics [12]

Relationship to other models

E. System Dynamics [3,4,9]

Relationship to other models

F. Gaming Techniques [14]

III. Appropriateness Criteria for Model Choice [12]

A. Definition of Model Goals

1. Scope of goals

   Degree of specificity

2. Relation of model goals to goals of overall policy

B. Nature of Inherent Assumptions in the Model [21]

1. Degree to which model is considered value-free

2. Assumptions related to all other appropriateness criteria

C. Statement of Objectives (Potential)

D. Statement of Constraints (Limits)

E. Dynamic Nature of Model Results as a Function of All Operational Criteria [23]

F. Correlation Between Spatial and Temporal Resolution of Model and Horizon of Policymaker [22]

G. Level of Substantive Aggregation [17,23]

1. Model description

   a. Level of aggregation of individual state variables

   b. Geographical range

   c. Environmental range

   d. Relation of level of aggregation to required decision-making

H. Means of Optimization [12]

1. Analytic manipulation

2. Extensive simulation runs

   a. With screening models

   b. Without screening models

I. Inclusion of Real-World Factors [2]

1. Social forces

2. Political forces

J. Degree of Linkage Among Model Sybsystems [23]

1. (e.g.) Physical/Chemical processes linkages

2. Decoupling possible

K. Economic Feasibility of Model Development (Cost)

1. Preliminary preparation

2. Programming effort

3. Running time

4. Validation time

L. Degree of Involvement of Decision-Maker [22]

1. Degree to which decision-maker feels threatened by model existence

2. Degree to which decision-maker understands model limitations

3. Simplicity of model implementation by potential user

4. Extent of communication with modelers on team

5. Role of modeler's proprietary interest in model (modeler becomes advocate)

6. Linkage between modeler's staff and staff of decision-maker

7. Degree to which common definitions are shared

   a. Prediction

   b. Validation

M. Degree to Which Effective Coordination of Interdisciplinary Team is Achieved [24]

1. Common languages of expression among team

   Common definitions of important modeling terms

   (1) prediction
   (2) verification

2. Common mathematical basis

3. Common understanding of economic factors

4. Common understanding of team goals

5. Agreement on spatial and temporal resolution

   Time-horizon of each specialist understood by all others

6. Existence of acceptable and competent peer review system

7. Psychological factors

   a. Team co-ordinator to determine correct motivational image (non-"brain-picker")

   b. Strive for "team image"--common purpose, mutual advantages and goals

   c. Degree to which mutual respect is generated among team members

N. Degree of Acceptance of Model by Scientific Community

O. Relation of Model to Other Management Areas

(e.g.) Monitoring

   a. Improve design criteria for network development

   b. Improve nature of data collected

   c. Reduce error levels

IV. Operational Criteria for Model Choice

A. Prototype Model Experimentation [11,23]

1. Formulation

2. Validation

   a. During model development

   b. After model is developed

   c. Emphasis on validation as a continuing process

B. Nature of Information Base

1. Scope

   Number of disciplines considered

2. Depth within discipline

C. Choice of Computer Language

D. Comparability with Existing Models

E. Model Validation Processes [11,13,23]

1. Potential for operational instability

2. Existence of systematic bias in data base (e.g. ambient pollutant levels)

3. Statistical limits on extrapolating the results of the model

4. Correlation with confidence level given by decision-maker

5. Sensitivity analysis

a. Using various data bases

b. Using other models with same objectives

F. Technical Feasibility

1. Computer capability

a. Data manipulation

b. Storage space

c. Storage time

2. Precision limitations

G. Nature of Personnel Needed to Operate Model

1. Quantity

2. Degree of training

H. Achievement of Results in Time Specified by Decision-maker [22]

V. <u>Models Available for Use vs. Appropriateness and Operational Criteria (matrix)</u>

VI. <u>Candidates for Modeling: Environmental and Social Systems</u>

A. Environmental Media

1. Air sheds [10]

2. Lakes

3. River basins [5]

4. Estuaries

5. Coastal ocean areas
6. Deep ocean areas
7. Soil types
8. Climate and weather processes

B. Terrestrial Ecosystems
   1. Forests
   2. Grasslands
   3. Deserts
   4. Arctic and Antarctic regions (Tundra?)
   5. Agricultural systems [6]

C. Aquatic Ecosystems [8,19]
   1. Freshwater
   2. Marine
   3. Estuarine

D. Resources [1]
   1. Energy fuels
      a. Reserves available vs. cost
      b. Consumption trends
      c. Technological changes
         (1) fuel use
         (2) fuel conservation
   2. Non-fuel minerals
      a. Reserves available vs. cost
      b. Consumption trends
      c. Technological changes
         (1) use
         (2) conservation
         (3) recycling

3. Other non-mineral resources

E. Behavior of Organisms and Populations [23]

1. Organism behavior

2. Species behavior

3. Population dynamics

4. Migration patterns

5. Genetic resources

F. Land Use [6,20]

1. Developed settlements

2. Preservation of non-developed areas (parkland, wilderness)

G. Human Living

1. Population density

2. Sanitation and water supply

   outbreak of diseases

3. Mortality rates for selected diseases

   a. Chemical pollutant-induced

   b. Radioactive-pollutant-induced

   c. Occupational

H. Economic Activity

1. Simulation of national economy

2. Capital expenditure models for pollution control

I. Aggregate National and Global Systems [2,7,9]

VII. Models Available for Use Vs. Systems to Study

For this topic we plot Topic II Items vs. Topic VI items in terms of:

A. Topics III and IV items, for comparison of advantages and disadvantages

B. State-of-the-art

C. Information sources for further detailed information (if our client so wishes)

MODELING

Bibliography

1. Baughmann, Martin L. Dynamic Energy System Modeling--Interfuel Competition. M.I.T. Energy Analysis and Planning Group. Sept. 1972.

2. Cole, H. S. D., et al. (editors) Models of Doom: A Critique of the Limits to Growth. Universe Books, New York. 1973.

3. Forrester, Jay W. Industrial Dynamics. The M.I.T. Press, Cambridge, Massachusetts. 1961.

4. Forrester, Jay W. Principles of Systems. Wright-Allen Press, Cambridge, Massachusetts. 1970.

5. Hamilton, H. R., et al. Systems Simulation for Regional Analysis: An Application to Reves-Basin Planning. The M.I.T. Press, Cambridge, Massachusetts. 1969.

6. Heady, E. O., et al. "National and Interregional Models of Water Demand, Land Use, and Agricultural Policies." Water Resources Research. vol. 9. no. 4. August 1973. pp. 777-791.

7. "I.I.A.S.A. Ponders Initial Research Program." Public Science. March 1973. pp. 7-14.

8. James, I. C.; Bower, B. T.; and Matalas, N. C. "Relative Importance of Variables in Water Resources Planning." Water Resources Research. vol. 5. no. 6. Dec. 1969. pp. 1165-1173.

9. Meadows, Donella H., et al. The Limits to Growth. Universe Books, New York. 1972.

10. National Industrial Pollution Control Council. Mathematical Models for Air Pollution Control Policy Decision-Making. U.S. Government Printing Office, Washington, D.C. February 1971.

11. de Neufville, Richard and Marks, David (editors). Systems Planning and Design: Case Studies in Modelling, Optimization and Evaluation. (Draft Prepublication Edition) M.I.T. Department of Civil Engineering. January 1973.

12. de Neufville, Richard and Stafford, J. H. Systems Analysis for Engineers and Managers. McGraw-Hill, New York. 1971.

13. Raiffa, Howard. Decision Analysis: Introductory Lectures on Choices under Uncertainty. Addision-Wesley, Reading, Massachusetts. 1970.

14. Rogers, Peter. "A Game Theory Approach to the Problems of International River Basins." Water Resources Research. vol. 5. no. 4. August 1969. pp. 749-760.

15. Russell, Clifford S. and Spofford, Walter O. Jr. "A Quantitative Framework for Residuals Management Decisions." Reprinted from A. V. Kneese, and B. T. Bower, editors. Environmental Quality Analysis: Theory and Method in the Social Sciences. The Johns Hopkins Press, Baltimore, Maryland. Chapter 4.

16. Simmons, Donald M. Linear Programming for Operations Research. Holden-Day Inc., San Francisco. 1972.

17. Spofford, Walter O. "Total Environmental Quality Management Models." Models for Environmental Pollution Control. Ann Arbor Science Publishers, Inc. 1973.

18. U.S. Department of Health, Education, and Welfare, Report of the Task Force on Research Planning in Environmental Health Science, "Man's Health and the Environment--Some Research Needs." March 10, 1970. (See especially Chapter 10, "Technological Trends.")

19. U.S. Environmental Protection Agency, Office of Water Programs. "Systems Analysis for Water Quality Management Survey and Abstracts." Document #SD-1-09-71. U.S. Government Printing Office, Washington, D.C. September 1971.

20. University of California at Davis, Interdisciplinary Systems Group. Land Use, Energy Flow, and Decision Making in Human Society. Second Annual Report.

21. Fishman, George S. Concepts and Methods in Discrete Event Digital Simulation. Wiley and Sons, New York. 1973.

22. Manheim, Marvin L. Model-Building and Decision-Making. M.I.T. Research Report R62-10. Civil Engineering Systems Laboratory.

23. Goodall, David W. "Problems of Scale and Detail in Ecological Modeling." Journal of Environmental Management. 2. 1974. pp. 149-157.

24. Discussion with Professor Brian W. Mar, Department of Civil Engineering, University of Washington, Seattle, Washington.

MONITORING

A fundamental responsibility of proper environmental management is the systematic accumulation of information about significant physical and social system parameters. Adequate monitoring capability can respond to a manager's need for data by providing comprehensive facts in a prompt manner within a tolerable degree of accuracy. Moreover, linkages among various monitoring systems allow decision-making on a broad level of action to be executed with the confidence that all critical segments of a given problem have been analyzed and reported.

Results from monitoring systems can provide the key factor in many problem assessment areas of environmental management, from initial problem definition (e.g. the carrying capacity of a given land area for extensive industrial development) to continuing post-project analysis (e.g. recording the actual migration of population to a given area after development has been initiated). Knowledge of the manner in which such information can be collected, and the limitations of such efforts, gives the environmental manager an initial basis for judgement as to the overall feasibility of assessment for any given undertaking. Another dimension of the utility of monitoring systems is related to the role these systems perform as part of an overall regulatory mechanism. The existence of certain monitoring techniques and systems may be the principal motivation for the enactment of standard-setting legislation.

It is apparent that monitoring processes can serve a crucial role in providing adequate assessment capability for a given issue. The outline presented here for this subject area is a first step towards providing the environmental manager with both an understanding of the various available systems and a set of criteria for use in choosing a system within a given situation. With these concepts in hand it is felt that the integration of monitoring efforts with other related tasks will be performed with a better measure of efficiency.

MONITORING

General Outline

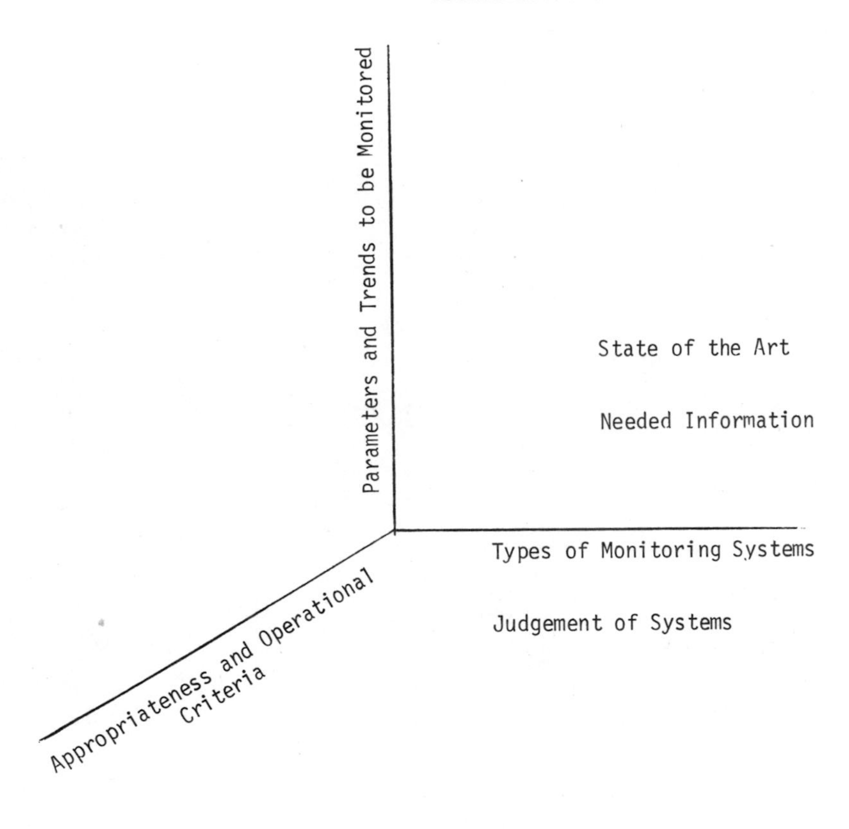

MONITORING

Detailed Outline

I. The Nature of Monitoring Systems

A. Definitions of Monitoring

1. UN Definition[11]: Monitoring--a system of continued observation, measurement, and evaluation for defined purposes

2. SCEP Definition[9]: Monitoring--systematic observations of parameters related to a specific problem, designed to provide information on the characteristics of the problem and their changes with time

3. Other definitions

B. Basic Assumptions of Monitoring Systems

1. Statistical assumptions

a. Optimal level of monitoring

b. Optimal location of monitoring

2. Assumptions on approach to monitoring

a. Types of species to monitor vs. objective of monitoring program

b. Trade-offs to implement the types of systems available (e.g. biological vs. chemical vs. physical)

3. Assumptions on analytical method used

4. Basis for assumptions

Reliance on current scientific information

II. Types of Monitoring Systems

A. Monitoring of Physical and Chemical Systems [2,3]

1. Dynamics of physical systems

a. Air transport processes

b. Hydrologic processes

c. Climatic processes

2. Chemical pollutant environmental parameters [10]

   a. Air pollutants--NOx, SOx, etc.

   b. Water pollutants

      (1) organic chemicals
      (2) nitrates
      (3) phosphates

   c. Microbiological contaminants

   d. Radioactive material

B. Biological Monitoring Systems

   1. Indicator species

   2. Population structure

   3. Population density

   4. Population dynamics as an indication of population health

C. Economic Monitoring

   1. Size of economy

   2. Type of economy

      a. Industrial

      b. Commercial

      c. Service

   3. Employment statistics

D. Social System Monitoring [19]

E. Gathering of Information on Social Preference [19]

III. <u>Technical Approaches in Monitoring Systems</u>

A. Sequential Categorization [5]

   1. Systems of warning

   2. Predictive systems

   3. After-the-fact assessment systems

B. Spatial-Sequential Categorization

1. Base-line monitoring

a. Requirement of pollution free or naturally wild area

b. Provision for early warning system

c. Criteria for transfer from base-line to ambient monitoring

2. Ambient monitoring

a. Purpose--a fixed network operation to determine environmental trends, such as pathways and fates of pollutants

b. Linkage with early warning system

3. Source monitoring

a. Stationary sources (pollutants and resources)

b. Mobile sources (pollutants only)

IV. Appropriateness Criteria for Choice of Monitoring System

A. Relationship of System to Overall Objective [4,6]

1. Policy objective

2. Legal Mandate

B. Dynamic Nature of System

C. Level of Aggregation

1. Geographical

2. Environmental

D. Nature of User Interaction

1. Active (User can manipulate)

2. Passive (User only receives data)

E. Economic Feasibility (Cost)

1. Planning costs

2. Implementation costs

F. Capacity to Transfer Knowledge to Decision-making Mechanism

1. Mechanical means

2. Non-machine interface

V. Operational Criteria for Choice of Monitoring System

A. Accuracy of System Results [4,17,22]

1. Statistical validation

a. Network operation

b. Received information

2. Sensitivity analysis performed on:

a. System size

b. Data specimen quality

B. Continuity of System in Time

C. Compatibility with Existing Systems and Programs [17]

D. Technical Constraints [17]

1. Present

2. Future forecast

E. Availability of Appropriate Target Area

F. Personnel Requirements [17]

1. Quantity

2. Literacy

3. Necessary training

VI. Monitoring Systems vs. Appropriateness and Operation Criteria (matrix)**

VII. Criteria for Cost-Effective Monitoring Programs for Developing Countries [20]

*A. Minimum Requirement (Examples)

1. Bacteria in water supplies monitoring for coliform count to avoid hepatitis

2. Data collection on dumping practices near water supplies

3. Monitoring of food-stuffs to avoid poisoning

---

*It is assumed that developed countries have these programs already.
**Matrix of items in Topic II versus items in Topics IV and V.

*B. Establishment of Links Between Principal Environmental Effects in the Country and Nature of Monitoring System

*C. Monitoring of Higher Trophic Level Ecosystems for Long-lived Pollutants (e.g. metals, pesticides)

D. Gradual Transformation to Standard Monitoring System Categories Given in Topic III

VIII. Uses of Monitoring Systems

A. Research

1. New analytical methods

2. New information processing techniques

B. Information Exchange [7,12,13,21]

1. e.g. Earthwatch Data and Information Centers

2. Use of UNEP information exchange systems

C. Testing and Design of Theoretical Models [14,18]

1. e.g. GARP Weather Modeling

2. e.g. pollutant pathway and fate modeling for ecological models

D. Enforcement of Regulations

Case preparation monitoring--a temporary network operation to generate data for proving causality, implemented primarily by government (and perhaps citizens) to establish legal nature of adequate data

E. Hazard Assessment

e.g. Worldwide Standardized Seismic Network provides observations that contribute to seismic risk maps

F. Evaluation and Review

1. Basis for setting criteria and standards

2. Cost assessment of programs

IX. Parameters and Trends that Can Be Monitored

A. Basic Environmental Media Behavior [6,10,15]

1. Air sheds

---

*It is assumed that developed countries have these programs already.

2. Lakes
3. River basins
4. Estuaries
5. Coastal ocean areas
6. Deep ocean areas
7. Soil types
8. Climatic and weather processes

B. Terrestrial Ecosystems

1. Forests
2. Tundras
3. Grasslands
4. Deserts
5. Agricultural systems

C. Aquatic Ecosystems

1. Marine
2. Freshwater

D. Resources

1. Energy fuels
    a. Reserves available as a function of cost
    b. Consumption trends
    c. Technologies for development and conservation
2. Non-fuel minerals
    a. Reserves available as a function of cost
    b. Consumption trends
    c. Technologies for development and conservation

E. Animal Wildlife

1. Population dynamics

2. Migration patterns

3. Genetic resources

F. Land Use

1. Developed settlements

2. Preservation of non-developed area (parkland)

G. Human Living

1. Sanitation and water supply system

    Outbreaks of disease

2. Population density

3. Mortality rates for selected diseases

    a. Chemical pollutant--induced

    b. Radioactive pollutant--induced

    c. Occupational

MONITORING

Bibliography

1. American Chemical Society. Cleaning Our Environment: The Chemical Basis for Action. American Chemical Society, Washington, D.C. 1969.

2. Atkinson, Arthur, and Gaines, Richard S. (eds.) Development of Air Quality Standards. Charles E. Merrill Publishing Co., Columbus, Ohio. 1970.

3. Baseline Studies of Pollutants in the Marine Environment and Research Recommendations: the IDQE Baseline Conference, May 24-26, 1972. New York. 1972.

4. Mitre Corporation. The Environment: A Systems Approach with Emphasis on Monitoring. Document M70-50. October 1970.

5. National Aeronautics and Space Administration. Remote Measurement of Pollution. Report of a Working Group sponsored by the National Aeronautics and Space Administration, arranged and administered by Langley Research Center, and convened at Norfolk, Virginia, August 16-20, 1971. Washington, D.C. 1971.

6. Sayers, William T. "Water Quality Surveillance: the Federal-State Network." Environmental Science and Technology. vol. 5. no. 2. February 1971. pp. 114-119.

7. Smithsonian Institution. National and International Environmental Monitoring Activities: A Directory. Cambridge, Massachusetts. October 1970.

8. Study of Man's Impact on Climate. Inadvertent Climate Modification. M.I.T. Press. 1971.

9. Study of Critical Environmental Problems. Man's Impact on the Global Environment: Assessment and Recommendations for Action. M.I.T. Press. 1970. Read p. 167-177.

10. Thomas, W. A.; Babcock, L. R. Jr.; and Shults, W. D. Oak Ridge Air Quality Index. ORNL pub. #ORNL-NSF-EP-8. September 1971.

11. United Nations Conference on the Human Environment, Intergovernmental Working Group on Monitoring or Surveillance. Report of the First Session. Document # A/CONG.48/IWGM.I/8. August 27, 1971.

12. United Nations Environmental Programme. Action Plan for the Human Environment: Programme Development and Priorities (Report of the Executive Director). Document # UNEP/GC/5. April 2, 1973.

13. U.S. Council on Environmental Quality. The Federal Environmental Monitoring Directory. U.S. Government Printing Office, Washington. May 1973.

14. U.S. Department of Commerce, *et al*. *World Weather Program: Plan for Fiscal Year 1974*. U.S. Government Printing Office, Washington. 1973.

15. U.S. Environmental Protection Agency. *Nationwide Air Pollution Emission Trends, 1940-1970*. January 1973.

16. U.S. Environmental Protection Agency, Office of Research and Development. *Environmental Protection Agency's Monitoring Programs*. Washington. August 1973.

17. U.S. Environmental Protection Agency, Office of Monitoring, Quality Assurance Division. *Development of Agency-Wide Quality Control Program*. February 13, 1973.

18. U.S. Interagency Committee for Global Environmental Monitoring. "Proposal for Initial Implementation of Global Environmental Monitoring." December 17, 1973.

19. U.S. Environmental Protection Agency, Office of Research and Monitoring, Environmental Studies Divisions. *The Quality of Life Concept: A Potential New Tool for Decision-makers*. Report of a symposium held August 29-31, 1972.

20. This information came primarily from an interview with Mr. Kevin Mullen, EPA Office of Research and Development.

21. Cition, Robert. *A Plan for the Implementation of the United Nations Global Environmental Monitoring Systems (GEMS), 1974-1978*. Smithsonian Institution. May 1974.

22. Mage, D. T. "Air Quality Monitoring: Playing Blind Man's Bluff with Pollution." *New Engineer*. June 1974. pp. 23-27.

23. Schafer, C. J. and Lailas, N. "Coping with Discharge Regulations." *Environmental Science & Technology*. vol. 8. no. 10. October 1974. pp. 903-6.

GROWTH AND ITS IMPLICATIONS FOR THE FUTURE

Development of a comprehensive approach to environmental management concepts in an educational format requires some exposure to the underlying perspectives of broad-scale social behavior and trends. In the area of environmental affairs, the processes with which we are most interested are those that describe the past and present physical development modes of society and the various social motivations for these modes. This is because physical development modes provide the setting within which values are established and environmental disruptions result, and are perceived.

The most common mode of physical development by nation-states in the world today is a growth mode. This is characterized principally by attempts to expand the national industrial and economic base. Population growth often naturally accompanies this economic growth, and can fuel a need for further growth. The classical rationale for such expansion is that the quality of life increases as growth continues.

But today the traditional industrial growth mode is being disturbed by disruptive "side-effects" and also being attacked philosophically as not providing an adequate raison d'etre for daily life. Principal among these "side-effects" are the environmental implications of further growth and we must understand these in order to manage. At the same time, developing countries must be provided with the means for development consistent with an actual and not falsely perceived improvement in the quality of life. Understanding the benefits and hazards of growth is the first step in plotting viable futures for society.

GROWTH AND ITS IMPLICATIONS FOR THE FUTURE

General Outline

Basic Parameters Related to Growth

Benefits and Hazards of Growth

Evaluation of Adjustment Mechanisms

Public Policy Issues

Growth Parameters and Societal Indicators

GROWTH AND ITS IMPLICATIONS FOR THE FUTURE

Detailed Outline

I. Growth: Basic Parameters and Technical Considerations

A. Resource Availability [22,24]

1. Energy resources
2. Food resources
3. Land and water resources
4. Non-fuel mineral resources
5. Other resources

B. Environmental Pollution [17,18]

1. Air
2. Water
3. Other

C. Population Growth [11,16]

1. Levels and rates of increase of population
2. Population distribution

II. Benefits and Hazards of Growth

A. Benefits of Growth [14]

1. Development of the third world
2. Greater GNP
3. More employment
4. More modern conveniences
5. Greater growth in technology
6. Greater ability to cope with social problems
7. More welfare to the poverty-stricken
8. Greater opportunities to improve oneself

   Growth is akin to the Protestant Ethic (Horatio Alger myth, etc.)
9. Others

B. Hazards of Growth

1. Population crowding [11,12,16]

2. Energy problems

   a. Impact of energy in perturbing the atmosphere to produce climatic change

   b. Environmental degradation

   c. Growing shortages in fuel supplies

   d. Role of energy on social, economic and political issues

   e. Dangers of oil and gas substitutes (i.e., nuclear energy, LNG)

3. Exhaustion of non-fuel minerals [22,24]

4. Food shortages

   a. Demand for food is increasing:

      due to population increase and demand for animal protein

   b. Supply of grain:

      (1) supply of arable land
      (2) what is the limit to photo-synthetic conversion of sunlight to edible calories in terms of calories per hectare?
      (3) world consequences if U.S. reduces exports of wheat due to:

         (a) the rising American demand
         (b) disruption of U.S. supply due to natural causes , etc.

5. Climatic change [17]

   a. Global instability (i.e., melting of the ice caps)

   b. Regional instability (i.e., drought)

6. Increased release of toxic substances affecting:

   a. Oceans

   b. Fresh water

   c. Atmosphere

   d. Climate

7. Interference with solar radiation

What are the factors controlling the quantitative spectrum of solar radiation reaching the earth's surface? What are the principal means by which we might purposely or inadvertently interfere with these factors and which are the limits to such interference without producing changes in the spectrum that would have major effects on life?

8. Human stress

In urban settlements

(1) health hazards (psychological and physiological)
(2) traffic congestion
(3) depersonalization of lifestyle and work

9. The endangering of various wildlife species

10. The endangering of wilderness

11. The problems associated with maintaining a harmonious world community [2]

a. Competition for scarce resources

b. Problems associated with pollution crossing borders

c. Growing gap between the Industrialized World and the Third World

d. Terrorist's activities for leverage

12. Other hazards (particularly those subsystems of the world and national scene which can be found by completing the matrix in topic V)

III. Evaluation of Adjustment Mechanisms

A. Market Mechanisms [4,14]

B. Technology

1. "Technical-Fix-Certain" (application of alread-existing technology)

2. Technological development and innovation--The "Technical-Fix-Hoped-For"

C. Constraints Resulting from Institutions, Value Structures, Cultural Attitudes, etc. [5,7]

1. Unemployment and the role of work in society [21]

2. Inflation

3. World monetary system

4. Other

D. Present and Suggested Policy Controls [19]

1. Present controls:

a. Land use controls [20]

b. Fiscal policy

c. Monetary policy and the Federal reserve

d. Other

2. Suggested controls:

a. "Grandchild impact statements"

b. "Communitarian" constraints upon the excesses of individualism (value constraints--value change)

c. Increased activism--citizen participation

3. Levers available in other countries

IV. <u>Public Policy Issues</u>

A. International Issues

1. Growth, no-growth and the development of the Third World

2. International distribution of the costs and benefits of growth and no-growth

3. R.I.C.--"resource-inspired conflict"

B. National Issues [8,19,23]

1. Maintenance of full employment

2. Social stresses accompanying a steady state economy or an economic decline

a. Old stresses that were masked by growth or neutralized by the expectation of sharing in the benefits of growth

b. New stresses created by a steady state economy or by an economic decline

3. Environmental protection [15,18]

C. Provincial and Local Issues

1. Provincial efforts to control growth--methods and impact

2. The local no-growth policies and methods--purposes, effects and legal challenges

3. Distribution of costs and benefits of provincial-and-local efforts to manage growth

4. Land use controls as a method for managing growth

5. Is the assumed "right of mobility" an absolute right?

D. How Policy Issues Derive from Adjustment Mechanisms

V. Social Indicators vs. Growth Parameters

Growth Parameters / Societal Indicators	Population	Food	Economic Development	Pollution	Resource Usage	Land Use	Other	Total
Social and Cultural								
Population Density								
Welfare								
Education								
Institutional Diversity								
Mobility								
Recreation								
Liesure								
(etc.)								
Economic								
GNP								
Employment								
Resource Usage								
Resource Availability								
Agricultural Sector								
Industrial Sector								
Free Market Viability								
Balance of Payments								
(etc.)								
Political								
Domestic								
International (incl. "Third World")								
(etc.)								
Physical Systems								
Human Health								
Wildlife								
Plants & Vegetation								
Aquatic Creatures								
Air Sheds								
Water Basins								
Land Use								
Climate								
(etc.)								
Total								

Each element in this matrix would represent the perceived effect of the "growth parameter" on the "social indicator"

******

The nature and degree of such an effect could be recorded either as quantitative or qualitative. Total aggregate effects would then be subjectively interpreted, in relation to desired societal goals.

******

GROWTH AND ITS IMPLICATIONS FOR THE FUTURE

Bibliography

1. Barnett, H. J. and Chandler, Morse. Scarcity and Growth. Johns Hopkins Press, Baltimore, Md. 1963.

2. Brown, Lester. World Without Borders. Vintage Books, New York. 1972.

3. Cole, Freeman, Jahoda and Pavitt (eds.). Models of Doom. Universe, New York. 1973.

4. Daly, Herman. Toward a Steady-State Economy. Freeman Press, San Francisco. 1973.

5. Gabor, Dennis. The Mature Society. Secker & Warburg, London. 1972.

6. Hardin, Garrett. Exploring New Ethics for Survival. Viking Press, New York. 1972.

7. Geiger, Theodore. The Fortune of the West. Indiana Press. 1973.

8. Goldsmith (ed.). Blueprint for Survival. The Ecologist. Jan. 1972.

9. Holdren, John P. and Ehrlich, Paul. Global Ecology. New York. 1971.

10. Forrester, Jay. World Dynamics. Wright-Allen Press, Cambridge, Mass. 1971.

11. Hardin, Garrett. "The Tragedy of the Commons." Science. vol. 162. pp. 1243-1248.

12. Meadows, D., et al. The Limits to Growth. Potomac Associates, Washington, D.C. 1972.

13. Toffler, Alvin. Future Shock. Random House. 1970.

14. Wallich, Henry L. "How to Live with Economic Growth." Fortune. October 1972.

15. Ward, B. and Dubos, R. Only One Earth. Norton, New York. 1972.

16. Population and the American Future. The Report of the Commission on Population and the American Future. USGPO, Washington. 1972.

17. Study of Man's Impact on Climate. Inadvertent Climate Modification. M.I.T. Press, Cambridge, Mass. 1971.

18. Study of Critical Environmental Problems. Man's Impact on the Global Environment. M.I.T. Press, Cambridge, Mass. 1971.

19. Growth and Its Implications for the Future. Hearings for the Subcommittee on Fisheries and Wildlife, Conservation and the Environment of the Committee on Merchant Marine and Fisheries, U.S. House of Representatives, Parts 1, 2, and 3. USGPO, Washington. 1973.

20. Rockefeller Brothers Fund. The Use of Land. Task Force Report. Crowell Publishing Co., New York. 1973.

21. Report to the Secretary of the Department of Health, Education, and Welfare, U.S. Government. Work in America. M.I.T. Press, Cambridge, Mass. 1973.

22. National Academy of Sciences/National Research Council, Committee on Resources and Man. Resources and Man. Freeman and Company, San Francisco. 1969.

23. Choucri, N.; Laird, M.; and Meadows, D. L. "Resource Scarcity and Foreign Policy: A Simulation Model of International Conflict." M.I.T. Center for International Studies. Publication C/72-9. March 1972.

24. Brooks, D. B. and Andrews, P. W. "Mineral Resources, Economic Growth, and World Population." Science. vol. 185. July 5, 1974. pp. 13-19.

ECONOMICS OF EXTERNALITIES

The issues and problems of environmental management can be very diverse, ranging from almost purely abstract concerns such as how to define the beauty of a parkland to extremely sophisticated analytical considerations such as air transport models for chemical particulate matter. However, in most circumstances the environmental manager will be confronted with problems that have significant dimensions in the economic sphere, since almost any proposal for environmental policy analysis requires the considerations of economic and technical implications.

An understanding of the fundamentals of economic systems is essential for the environmental manager and that knowledge is a prerequisite to the information in this subject area. What is often needed however is to focus that fundamental knowledge within a framework of economic principles related to classical theory, the theory of welfare economics, and the various interpretations of the causes of environmental "externalities" as they are explained in economic theory.

As we developed various concepts on the basis of the underlying economic theories available, we found that the most efficient means of packaging this information would be in the form of a discussion of the types of environmental "externalities" produced and how the economic impacts of these residuals could be forecast. After this, the principal issue area to consider is how economic costs and benefits can be assessed in analyzing the role of environmental pollution and depletion of resources. Certain specific questions of an economic nature arise in the private marketplace and these must be treated as well. Finally, much of the information discussed in this subject area will serve us well in attempts to elaborate an environmental impact assessment methodology.

ECONOMICS

General Outline

Economic Perspectives and Principles

Technical Perspectives of Residuals Generation

Economic Analysis Procedures

Important Issues

THE ECONOMICS OF EXTERNALITIES

Detailed Outline

I. Economics Perspective and Principles [8,9]

  A. Classical Economic Principles

  B. Welfare Economics [19]

  C. Concepts of Social and Private Costs [19]

  D. The Definition of Externalities

    1. Pollution as "generic congestion" [20]

    2. Others [11]

  E. The Deterioration of Common Property Resources [10]

  F. Property Rights and Boundaries

  G. Concept of Public Good [3]

II. Technical Perspective of Residuals Generation

  A. Sources of Residual Generation [9]

    1. Production sources vs. combustion sources vs. disposal sources

    2. Residuals associated with product type

      a. Non-fuel mineral industries

      b. Energy industries

      c. Other chemical processing

  B. Effects of Residuals on the Environment [7,12,13,18]

    1. Effects within the firm

      a. Worker health and efficiency

      b. Working environment

      c. Plant and equipment

2. Effects on the external environment

a. Human health

(1) physical
(2) psychological

b. Wildlife

c. Environmental systems

(1) air
(2) water
(3) land

d. Aesthetics and recreational effects

III. <u>Economic Analysis Procedures</u>

A. Classification of Costs [1,2,6,15,22]

1. Direct investment costs of project

2. Damage costs

a. Damage to health

b. Damage to materials

c. Destruction of ecosystems

d. Loss of aesthetic, recreational and other amenities

e. Intangible costs and their quantification

3. Avoidance costs

Difficulties in measurement

4. Transaction costs

a. Costs to achieve environmental goals and standards

(1) research
(2) development
(3) planning
(4) monitoring
(5) enforcement

b. Costs in preparing environmental impact statements

5. Abatement costs in managing activities

    a. Expenditure trends

        (1) private sector
        (2) public sector

    b. Estimating pollution control expenditures

    c. Impact of pollution control on the national economy

    d. Distribution of environmental expenditures

6. Social costs [5]

B. Classification of Benefits [13]

(examples: flood control benefits, land reclamation benefits; national income; regional income; income distribution)

C. Nature of Benefits and Costs [15,21]

1. Direct vs. Indirect

2. Internal vs. External

3. Tangible vs. Intangible

D. Specific types of Benefits and Costs

(Benefits--matrix of subtopic B vs. subtopic C)

(Costs--matrix of subtopic A vs. subtopic C)

E. Techniques and Problems of Economic Analysis [14,15]

1. Important techniques in analysis

    a. Benefit/cost ratio criterion

    b. Net benefits criterion

    c. Internal rate of return criterion

    d. Economic utility theory

2. Special problems

    a. Selection of a discount rate for project analysis

    b. Treatment of risk and uncertainty

    c. Effects of inflation

IV. Issues in the Management of the Firm that Relate to Environmental Controls and their Economic Implications [14,15]

A. Recycling Wastes

B. Product Switching

C. Mergers and/or Combinations of Enterprises for Pollution Control

D. Defining "Best Practicable" Control Technology

E. Markets

F. Labor Supply

G. Intra- and Inter-State Commerce [17]

H. Technical and Financial Assistance Schemes

I. Differences in Controls and their Effects for Large Companies vs. Small Companies

J. Profit and Social Responsibilities

K. Nature of Government Tactics for Pollution Controls* (e.g., Taxes vs. Subsidies, etc. [17])

L. Private Financing Alternatives for Pollution Control Expenditures

1. Stock divestiture (selling of shares)

2. Bank loans

3. Capital funds allocation

4. Price raise in product

5. Reduce production

M. Models Which Internalize Externalities [16]

1. "Materials Balance" approach

2. Others **

---

*cross reference with the relevant topics of the subject area on "Administrative Processes"

**cross references to the relevant topics in the subject area on "Modeling"

V. Questions of General Interest to the Firm

A. What will be the Trend of Environmental Law and Regulation? [17]

B. What Types of Effluent Regulatory Policies are Most Advantageous for the Firm? [13]

C. What is the Optimum Investment Schedule for Pollution Abatement Equipment?

D. What will be the Public Attitude Toward Corporate Responsibility in the Future? [4]

E. What Types of Management Personnel will be Needed to Tackle these Issues? [14]

THE ECONOMICS OF EXTERNALITIES

Bibliography

1. Ayres, R. V. and Kneese, A. V. "Production, Consumption, and Externalities." American Economic Review. June 1969. p. 282 ff.

2. Boulding, Kenneth. "The Economics of the Coming Spaceship Earth." Environmental Quality in a Growing Economy. Johns Hopkins Press, Baltimore. 1966. Published for Resources for the Future, Inc.

3. Buchanan, James, and Kafoglis, Milton. "A Note on Public Goods Supply." American Economic Review. July 1963. vol. 53. p. 403-410.

4. Committee on Economic Development. Social Responsibilities of Business Corporations. New York, New York. June 1971.

5. Coase, Ronald. "The Problem of Social Cost." Journal of Law and Economics. October 1960.

6. Daly, Herman. Toward a Stead-State Economy. W. H. Freeman & Co., San Francisco. 1973.

7. Daly, Herman. "Economics as a Life Science." Journal of Political Economy. vol. 76. no. 3. pp. 392-406.

8. Georgescu-Goegen, N. "The Entropy Law and the Economic Problem." University of Alabama "Distinguished Lecture" Series. no. 1. 1971.

9. Goldman, Marshall I. Ecology and Economics: Controlling Pollution in the '70's. Prentice-Hall, Inc., Englewood Cliffs, N.J. 1972.

10. Hardin, Garrett. "The Tragedy of the Commons." Science. vol. 162. pp. 1243-1248.

11. Heller, Walter W. "Coming to Terms with Growth and the Environment." In Schurr (ed.) Energy, Economic Growth, and the Environment. Johns Hopkins Press, Baltimore.

12. Kneese, A. V. Water Pollution: Economic Aspects and Research Needs. Resources for the Future, Inc. Johns Hopkins Press, Baltimore. 1962.

13. Kneese, A. V., and Bower, B. T. Managing Water Quality: Economics, Technology, and Institutions. Resources for the Future, Inc. Johns Hopkins Press, Baltimore. 1971.

14. Kneese, A. V. "Management Science, Economics, and Environmental Science." Management Science. vol. 19. no. 10. June 1973.

15. Kneese, A. V., and Bower, B. T. (eds.) Environmental Quality Analysis. Johns Hopkins Press, Baltimore. 1972.

16. Kneese, A. V.; Ayres, R. U.; and D'Arge, R. Economics and the Environment: A Materials Balance Approach. Johns Hopkins Press, Baltimore. 1972.

17. Kneese, A. V., et al. The Political Economy and Environmental Management. Wiley and Sons, Inc.

18. Mishan, E. J. "Reflections on Recent Developments in the Concept of External Effects." Canadian Journal of Economics and Political Science. February 1965.

19. Pigou, A. C. Economics of Welfare. MacMillan Co., New York. 1932.

20. Rothenburg, Jerome. "The Economics of Congestion and Pollution: An Integrated View." American Economic Review Papers and Proceedings. May 19, 1970.

21. Turvey, Ralph. "On Divergencies Between Social Cost and Private Cost." Economica. vol. 30. 1963. p. 309.

22. Williams, B. R. "Economics in Unwanted Places." Economic Journal. March 1965.

ENVIRONMENTAL LAW

The emerging professional practice of environmental management requires an understanding of the historical perspective of legal norms and the nature of modern legal approaches to environmental issues and problems. With respect to protecting environmental values, old concepts of common law including nuisance, tresspass, and negligence concepts are not sufficient today. Such doctrines were formulated during an expansionary ethos in Western society, and often reflect the view that resources are not exhaustible. The same reasoning could be applied to procedural law; used historically for settling disputes, it makes no provision for guaranteeing rights to a livable environment.

The reason for a subject area in environmental law focuses on an examination of the functional value of the law to meet societal needs. Law can be considered as a principal means for social regulation in attempting to formalize social values into a rational code of principles. Different legal systems have arisen as a result of modifying values and principles. Law usually not only defines substantive codes of conduct but also delineates the roles of individuals and groups in society who police the prevailing codes.

As a formal system of rules, laws can be enacted on many organizational levels, any or all of which may be in the manager's domain. The manager must be familiar with the substantive content of laws and the manner in which they specify administrative actions for enforcing precedent, setting standards, and regulating actions. In order to enhance understanding as to the interaction of legal systems and to plan the efficient use of resources in legal encounters it is necessary to examine the broader issues of how laws are formulated in any given political system and what implicit roles are available in seeking legal changes.

It is felt that this particular perspective will provide the environmental manager with a valuable background because it encompasses issues necessary for a comprehensive understanding of the workings of the legal structure in different organizational modes.

ENVIRONMENTAL LAW

General Outline

Values as a Basis for Law

Different Legal Systems

Controlling Activities and Actions

ENVIRONMENTAL LAW

Detailed Outline

I. Values as a Basis for Environmental Law [28,45]

A. Protection of Human Life Against Specific Environmental Risks [10]

B. Protection Against Environmental Risks Caused by Human Interaction With Nature

C. Recognition of Long-term Impact

D. Preservation of National and Natural Heritage for Future Generations

E. *Recognition of Public Goods and Amenities [29]

II. Approaches to Environmental Law in Different Legal Systems: A Comparison of Similarities and Differences Utilizing Case Studies [11,16,22,41]

A. National Legal Systems [8]

1. Common law
2. Civil law
3. Socialist law
4. Eastern law
5. Muslim law
6. African law [36]

B. Centralized and Decentralized Law

1. Federal systems vs. Unitary systems

a. Federal systems
e.g. United States
Canada
Australia
India
U.S.S.R.

b. Unitary systems
e.g. France

---

*Other subject areas that discuss this topic are "Values and Perception" and "Economics."

C. International Law

1. Types of activities and actions vs. jurisdictional claim [25,26]

a. Types of activities

(1) pollution of the oceans;
shipping;
river pollution;
ocean dumping;
marine pollution

(2) conservation of natural resources, [24]
endangered species, and
natural resources depletion [44]

(3) activities subject to international conventions and agreements

b. Jurisdictional claim

(1) national

(2) international
e.g. High Seas
Antarctica
Outer Space

III. <u>Types and Techniques of Legal Approaches: Legal Control of Environmental Activities</u> [3,15,27,40]

A. Activities and Pollutants

B. Environmental Law [6,14,17,18,37,38,42]

1. General [1,47,48]
<u>example</u> U.S.N.E.P.A., U.S. Water Resources Council

2. Air [21]

3. Water [2,12,20,21,23,35,46,49]

4. Solid and toxic wastes

5. Noise

6. Pesticides and other economic poisons [19]

7. Radiation

8. Land use [4,5]

C. Level of Jurisdiction

1. Federal

2. State

3. Local

D. Directives

1. Standards setting [50]

   a. Performance standards

   b. Quality standards

   c. Emission standards

   d. Labour standards

2. Permits

3. Enforcement procedures

4. Penalties

5. Responsibilities

6. Preparation of an E.I.S.

7. Procedures for registration

8. Requirements for labeling

9. Research and training programs

10. Inspections

11. Charges [13]

12. Licensing

13. Budgets and programs for pollution control and abatement and for activities which enhance or preserve the environment

E. Evaluation of Enforcing Directives

1. Evaluation of the enforcement of injuctions, fines, damages, penalties, etc.

2. Evaluation of enforcement programs

3. Relationship of the court with enforcement authorities

F. Environmental Challenge and Dispute Resolution [9,32,33,34]

1. The concept of an adversary system [7,39]

a. Actor and role interaction [43,51]

(1) interpersonal disputes
(2) government vs. private sector
(3) government regulating itself
(4) private vs. public
(5) environmental law organizations

b. The right to sue

2. Costs of environmental control

a. Court costs

b. Time in the court system

G. Selection of a Forum

H. Access of Information

1. Information accumulation and dissemination

2. Bureaucratic barriers

3. Adversary proceedings

ENVIRONMENTAL LAW

Bibliography

Cited References

The following references are those cited by the accompanying outline:

1. Anderson, Fred. NEPA in the Courts: A Legal Analysis of the National Environmental Policy Act. Johns Hopkins University Press, Baltimore, Maryland. 1973.

2. Baldwin, Frank B. III. (ed.) Legal Control of Water Pollution. University of California, Davis. 1969.

3. Baldwin, M. F., and Page, J. K. (eds.) Law and the Environment. Walker and Co., New York. 1970.

4. Berger, Curtis J. Land Ownership and Use: Statutes and Other Materials. Boston & Little, Brown. 1968.

5. Bosselman, F.; Callies; and Banta, J. The Taking Issue: An Analysis of the Constitutional Limits of Land Use Control. CEQ, Washington, D.C. July 1973.

6. Brocher, J. J. and Nestle, M. E. Environmental Law Handbook. Committee on Continuing Education of the Bar, State Board California, Berkeley. 1970.

7. Buckhorn, Robert. Nader: The People's Lawyer. Prentice-Hall, Englewood Cliffs, New Jersey. 1972.

8. Buxbaum, David C. (ed.) Traditional and Modern Legal Institutions in Asia and Africa. Leiden, E. J. Brill. 1967.

9. CEQ Legal Advisory Committee. Private Litigation on Environmental Protection. Washington, D. C. 1970.

10. Cooley, R. A. and Wandesforde-Smith (eds.) Congress and the Environment. University of Washington Press. 1970.

11. David R., and Brierly, J. E. Major Legal Systems in the World Today. New York and London. 1968.

12. Davis, P. N. "Theories of Water Pollution Litigation." Wisconsin Law Review. 1971.

13. Environmental Law Institute. Effluent Charges on Air and Water Pollution: A Conference Report on Law Related Studies. Washington, D.C. 1973.

14. E.P.A. Current Laws. U.S. Government Printing Office, Washington, D.C.

15. Grad, Frank P., et al. Environmental Control: Priorities, Policies, and the Law. Columbia University Press, New York. 1971.

16. Hargrove, John L. (ed.) Law, Institutions, and the Global Environment. Oceana Publications, Dobbs Ferry, New York. 1972.

17. Hassett, Charles M. (ed.) Environmental Law. Institute of Continuing Legal Education, Ann Arbor, Michigan. 1971.

18. Heath, M. S. Jr. Materials and Resource Law and Policy, Vol. 2. Institute of Government, University of North Carolina, Chapel Hill, North Carolina. 1972.

19. Henkin, H.; Merta, M.; and Staples, J. The Environment, the Establishment, and the Law. Houghton, Mifflin Co., Boston, Mass. 1971.

20. Holmes, Beatrice; Simons, George G.; and Ellis, Harold H. State Water-Rights Laws and Related Subjects: A Supplemental Bibliography, 1959 to Mid-1967. United States Government Printing Office, Washington, D.C. September 1972. 268 p.

21. Hurley, William. Environmental Legislation. C. D. Thomas Publ., Springfield, Illinois. 1971.

22. International Association of Legal Science. International Encyclopedia on Comparative Law. Executive Office Max Planck Institute, Hamburg, Germany.

23. Johnson, Carwin W., and Lewis, Susan H. (eds.) Contemporary Developments in Water Law. Center for Research in Water Resources, University of Texas, Austin. 1970. 177 p.

24. Johnston, Douglas M. The International Law of Fisheries. Yale University Press, New Haven. 1965. xxiv. 554 p.

25. Johnson, Douglas M., and Gold, Edgar. The Economic Zone in the Law of the Sea: Survey, Analysis, and Appraisal of Current Trends. The Law of the Sea Institute, Kingston, Rhode Island. 1973. 53 pp.

26. Jones, Erin Bain. Law of the Sea: Oceanic Resources. Southern Methodist University Press, Dallas. 1972. 162 pp.

27.. Kaina, M. (ed.) A Study on Pollution Law. Tokyo. 1969.

28. Krier, James E. Environmental Law and Policy. Bobbs-Merrill, Indianapolis. 1971.

29. Krier, J. E. The Pollution Problem and Legal Institutions: A Conceptual Overview. UCLA Law Review, Los Angeles, California. Reprint No. 73. 1971.

30. Law, Institutions and the Global Environment. Oceana Publications, Inc. New York.

31. Landau, Norman J. The Environmental Law Handbook. Ballantine Books, Inc. New York. 1971.

32. Lutz, R. E. and McCaffrey, S. C. "Standing on the Side of the Environment: A Statutory Prescription for Citizen Participation." Ecology Law Quarterly. Vol. 1. 1971. p. 561.

33. McDonald, J. B., and Conway, J. E. Environmental Litigation. Department of Law, University of Wisconsin, Madision. 1972.

34. McCarthy. "Recent Legal Developments in Environmental Defense." 19 Buff. L. Rev. 195. 1970.

35. Meyers, C. and Tarlock, D. Water Resource Management--A Coursebook in Law and Public Policy. The Foundation Press, Inc. Mincola, New York. 1971.

36. Mifsud, F. M. Customary Land Law in Africa. FAO Legislative Series No. 7, Rome. 1967.

37. Murphy, G. Laws Relating to the Protection of Environmental Quality. State of California, Dept. of General Service, Sacramento, California. 1970.

38. Mylraie, Gerald, and Flanagan, Hey. California Environmental Law--A Guide. Center for California Public Affairs, Claremont, California.

39. Private Litigation on Environmental Protection. CEQ Advisory Committee, Washington, D.C. 1970.

40. Reitz, Arnold Jr. Environmental Law. North American International, Washington. 1972.

41. Sand, P. Legal Systems for Environment Protection: Japan, Sweden, United States. F.A.O. (U.N.), Rome. 1972.

42. Sawai, H. Private Law Aspects of Pollution. Tokyo. 1969.

43. Sax, J. Defending the Environment, A Strategy for Citizen Action. Alfred A. Knopf, New York. 1971.

44. Tarlock, D. Problems of Land-Use Conflicts in International Law. University of Indiana.

45. Thomas, Fran. Law in Action: Legal Frontiers for Natural Resources Planning, the Work of Professor Jacob H. Beuscher. Land Economics, Madison, Wisconsin. 1972.

46. Trelease, Frank J. Cases and Materials on Water Law. St. Paul, Minnesota West. 1967.

47. U.S. Congress, House. National Environmental Policy Act of 1969. Conference Report to Accompany S. 1075. 91st Congress, 1st Session, December 17, 1969. U.S. Government Printing Office, Washington, D.C. 1969. 12 pp.

48. U.S. Congress, Senate. Compilation of Federal Laws Relating to Conservation and Development of Our Nation's Fish and Wildlife Resources, Environmental Quality, and Oceanography. U.S. Government Printing Office, Washington, D.C. 1972. 618 pp.

49. Walker, William R., and Cox, William E. Legal Aspects of Water Supply and Water Quality Storage. Virginia Polytechnic Institute, Water Resources Research Center, Blacksburg, Virginia. 1970. 235 pp.

50. Woolan, Michael J. The Process of Setting Safety Standards in the Courts, Congress, and Administrative Agencies. George Washington University Program of Policy Studies in Science, Washington, D.C.

51. Yannacone, V. J. and Cohen, B. S. Environmental Rights and Remedies. Rochester, New York. 1971.

General References

The following references are of general interest for this subject area:

American Association of Law Libraries. 64th Annual Meeting. Legal Bibliography of Current Social Problems. The Environment ... Law Library Journal 452.

Grad, Frank P. Environmental Law: Sources and Problems. Matthew Bender, New York. 1971.

Gray, Oscar S. Cases and Materials on Environmental Law. Bureau of National Affairs, Washington, D.C. 1971.

Oliensis, Samuel, et al. Electricity and the Environment: The Reform of Legal Institutions. Report Prepared for the Special Committee on Electric Power and the Environment, Association of the Bar of the City of New York, 1972. West, St. Paul, Minnesota. 1973. 322 pp.

Sherrod, H. F. (ed.) Environmental Law Review. Sage Hill, New York. 1971.

Sloan, I. J. Environment and the Law. Oceana Publications, Dobbs Ferry, New York. 1971.

Toubenfeld, Howard J. (ed.) Controlling the Weather: A Study of Law and Regulatory Processes. Dunellen, New York. 1970. 275 pp.

Trelease, Frank J.; Bloomenthal, H. S.; and Geraud, J. R. Cases and Materials on Natural Resources. St. Paul, Minnesota: West. 1965.

Turkin, E. W. Text, Cases, Problems on Legal Regulation of the Environment. St. Paul, Minnesota: West. 1972.

U.S. Congress, House. Committee on Merchant Marine and Fisheries. Subcommittee on Fisheries and Wildlife. Environmental Quality. Hearings before 91st Congress, 1st Session, May 7, 26, June 13, 20, 23, 26 and 27, 1969. U.S. Government Printing Office, Washington, D.C. 1969. 472 pp.

Winter, E. F. (ed.) Handbook on National Environmental Law Institutions. U.N. Conference on the Human Environment.

ADMINISTRATIVE PROCESSES

Environmental management can be considered as a process of discovering, analyzing, and making decisions about issues and problems concerning man's impact on the world and its resources. Within the usual societal context these actions are ascribed to both the individual in his daily life and the social institutions which transform resources into products.

For environmental management it is necessary to study the functions of administrative organizations from the perspective of how environmental control policies and strategies are formulated. Such an approach fits well with the other subject areas that have been developed since it details a type of actor/role interaction in the policy area that can be supplemented by consideration of other management dimensions. For example, explicitly considering the types of administrative controls available is one way of establishing how the range of monitoring and modeling strategies that are available (taken from development of the monitoring and modeling subject areas) can be applied to various environmental problems. Also, the economic dimension of environmental control is emphasized through a discussion of the rationale for implementing various controls. The role of environmental indicators and questions as to the availability and accuracy of relevant data need also to be considered from the viewpoint of the effectiveness of controls.

In considering environmental policy, any organization is forced with differences in the perception of environmental problems, the degree of commitment to values, and the ordering of policy priorities. An understanding of the available alternatives for control fills a needed gap in determining the options open to managerial action.

ADMINISTRATIVE PROCESSES

General Outline

Activities vs. Effects

Controls vs. Effectiveness

Organizations vs. Roles

ADMINISTRATIVE PROCESSES

Detailed Outline

I. Types of Activities vs. Effects

A. Types of Activities

1. Public

2. Private

B. Environmental Effects*

1. Pollution

2. Depletion

3. Degradation

II. Types of Administrative Controls vs. Effectiveness in the Administration of Controls [15,18,19]

A. Types of Administrative Controls

1. Regulation [3,30]

a. Standard setting [25]
e.g. B.O.D. standard

b. Tax regulation [27]
e.g. effluent charge

c. Quality control
e.g. "best practical technology"

2. Enforcement [12,35]

a. Penalties

b. Fines

c. Orders

3. Review

a. Ombudsman [14]

b. Central authority

c. Review by department

---

*For expansion of these topics, refer to the subject area on "Environmental Effects."

B. Effectiveness in the Administration of Controls [5,26]

1. Feasibility measures

a. % of variances granted
e.g., in standard setting

b. Number of court cases or non-payment of taxes

c. Man power, and available facilities

d. Variation in performance

III. Types of Organizations which Administer Controls vs. Organizational Roles

A. Types of Organizations which Administer Controls [5]

1. Government [21,22,31,32,33]

a. Local [8]
e.g., zoning board

b. Regional
e.g., river basin commission

c. National
e.g., Forest Service

d. Multi-national
e.g., International Joint Committee (Can.-U.S.)

e. International [4,7,17,24,38]
e.g., IMCO

2. Quasi-Government

a. Local
e.g., public service commission

b. Regional
e.g., state public utility commission

c. National
e.g., interstate commerce commission

d. International

3. Non-government
e.g., trade associations
labour unions
Underwriters Laboratories

B. Organizational Roles [1,11,13,23,28,29,34]

1. Regulatory
2. Assessment
3. Advisory
4. Promotive
5. Research
6. Development
7. Enforcement
8. Information
9. Conservation

Example 1: Type of Organization (National)

e.g. National Forest Service

Organizational Roles

e.g. Policy Formulation

Regulatory

Conservation

Development

Advisory

Research

Assessment

Financial Assistance

Example 2: Type of Organization (International)

e.g. U.N.E.P.

F.A.O.

U.N.E.S.C.O.

Organizational Roles

e.g. Administering specialized environment related agencies

Developing programs of actions

Advisory

Financial assistance

C. Institutional Environment [2,9,10,16,20,36,37]

1. Institutional constraints within the political environment

2. Constraints on the administrator

3. Organizational theory

ADMINISTRATIVE PROCESSES

Bibliography

Cited References

The following references are those cited by the accompanying outline:

1. Adams, Franklin S. "Environmental Decision Making: Retrospect and Prospect." Midwest Quarterly. 13. October 1971. pp. 101-116.

2. Anderson, Walt (ed.) Politics and Environment. Pacific Palisades. California: Goodyear. 1970. 362 pp.

3. Buggie, Frederick D., and Gurman, Richard. Toward Effective and Equitable Pollution Control Regulation. American Management Association, New York. 1972. 41 pp.

4. Black, C. E. and Falk, R. A. (ed.). The Structure of the International Environment. Vol. 4. Princeton University Press, Princeton, New Jersey.

5. Caldwell, Lynton K. "Authority and Responsibility for Environmental Administration." Annals of the American Academy of Political and Social Science. 389. May 1970. pp. 107-115.

6. Caldwell, Lynton K. "Public Policies and Environmental Values--Toward a Sounder Basis for Decisions." IUCN Bulletin--International Union for Conservation of Nature and Natural Resources, I. April-June 1964. pp. 1-2.

7. Caldwell, L. K. In Defense of Earth, International Protection of the Biosphere. Indiana University Press. 1972.

8. Cohen, Eleanor. Expanding the Environmental Responsibility of Local Government. Center for California Public Affairs, Claremont College Affiliate, Claremont, California. 1972.

9. Craik, Kenneth H. "The Environmental Dispositions of Environmental Decision-Makers." Annals of the American Academy of Political and Social Science. 389. May 1970. pp. 87-94.

10. Davies, J. Clarence. The Politics of Pollution. New York: Pegasus. 1970. 231 pp.

11. Degler, Stanley E., and Bloom, Sandara C. Federal Pollution Control Programs: Water, Air, and Solid Wastes. Bureau of National Affairs, Inc., Washington, D.C. 1971. III.

12. Delogu, Orlando E. "Effluent Charges: A Method of Enforcing Stream Standards." Maine Law Review. 19. 1967. pp. 29-48.

13. Deran, Elizabeth Y. Pollution Control: Perspectives on the Government Role. Tax Foundation, New York. 1971. 46 pp.

14. Ege, Karl J. "Enforcing Environmental Policy: The Environmental Ombudsman." Cornell Law Review. 56. May 1971. pp. 847-863.

15. Facht, Johan. Environmental Control in Sweden During the Seventies. Allmanna Forlaget, Stockholm, Sweden. 1971. 28 pp.

16. Gordon, Marsha and Morton. Environmental Management-Science and Politics. Allyn and Bacon, Inc., Boston, Mass. 1972.

17. "International Environment Management, Preliminary Thoughts." Natural Resource Journal. VII. no. 3. July 1971. 508 pp.

18. Jaffe, Louis L. "The Administrative Agency and Environmental Control." Buffalo Law Review. 20. Fall 1970. pp. 231-237.

19. Japan, Environment Agency. Air Pollution Control in Japan. Environment Agency, Tokyo. May 1972. 60 pp.

20. Kaufman, Herbert. The Forest Ranger: A Study in Administrative Behavior. The Johns Hopkins Press, Baltimore. 1960. 259 pp.

21. Kerbec, Matthew J. Your Government and the Environment Vol. II. Output Systems Corp., Arlington, Virginia. 1972.

22. Kohlmeier, Louis M. Jr. The Regulators-Watchdog Agencies and the Public Interest. Harper and Row Publishers, Inc., New York.

23. National Academy of Public Administration. Technology Assessment in State Government. National Academy of Public Administration, Washington, D.C. 1972. 71 pp.

24. Organisation for Economic Co-operation and Development. Governmental Responsibilities for the Application and Control of Technology in Relation to Man's Environment. Organisation for Economic Co-operation and Development (OECD), Paris. 1970. 37 pp.

25. Shiffman, Morris A. "The Use of Standards in the Administration of Environmental Pollution Control Programs." American Journal of Public Health. 60. February 1970. pp. 255-265.

26. Siegel, G. (ed.). Human Resource Management in Public Organization, A Systems Approach. University Publishers, Los Angeles, California. 1973.

27. United Nations Department of Economic and Social Affairs. Pollution Taxes. United Nations Department of Economic and Social Affairs, New York. 1971. 127 pp.

28. United Nations Department of Economic and Social Affairs. Public Administration Division. Administrative Aspects of Urbanization. (Based on a Comparative Study Carried Out with the Co-operation of the Institute of Public Administration of New York, and on the United Nations Workshop on the Administrative Aspects of Urbanization, held at the institute of Social Studies, at The Hague, Netherlands, November 11-20, 1968.) United Nations Public Administration Division, Department of Economic and Social Affairs, New York. 1970. 228 pp.

29. United Nations Department of Economic and Social Affairs. Public Administration Division. Organization and Administration of Environmental Programmes: With Special Reference to Recommendations of the United Nations Conference on the Human Environment, held at Stockholm, Sweden, June 5-16, 1972. United Nations Public Administration Division, Department of Economic and Social Affairs, New York.

30. United States Atomic Energy Commission. Regulatory Activities: 1972 Annual Report of the United States Atomic Energy Commission to Congress. United States Government Printing Office, Washington, D.C. 1973. 54 pp.

31. United States Advisory Commission on Intergovernmental Relations. The Quest for Environmental Quality, Federal and State Action, 1969-1970: Annotated Bibliography. Edited by Rachelle L. Stanfield, John J. Callahan, and Sandra Osbourn. United States Government Printing Office, Washington, D.C. 1971. 63 pp.

32. United States Congress. House. Committee on Government Operations. Subcommittee on Executive and Legislative Reorganization. Reorganization Plan No. 3 of 1970 (Environmental Protection Agency). Hearings of the Subcommittee to the Committee before the 91st Congress, 2d Session, July 22, 23 and August 4, 1970. United States Government Printing Office, Washington, D.C. 1970. 209 pp.

33. United States Congress. House. Committee on Merchant Marine and Fisheries. Subcommittee on Fisheries and Wildlife Conservation. Administration of the National Environmental Policy Act--1972. Hearings of the Subcommittee to the Committee before the 92nd Congress, 2d Session, February 17, 1972. United States Government Printing Office, Washington, D.C. 1972. 1887 pp. (Two Volumes)

34. United States Department of Commerce. The Challenge of the Environment: A Primer on EPA's Statutory Authority. United States Government Printing Office, Washington, D.C. 1972. 34 pp.

35. United States Department of Commerce. Office of the Comptroller General of the United States. Report to the Congress by the Comptroller General of the United States on the Water Pollution Abatement Program: Assessment of Federal and State Enforcement Efforts, March 23, 1972. United States Government Printing Office, Washington, D.C. 1972. 55 pp.

36. VanDersal, William R. "The Resource Administrator of Tomorrow." Journal of Soil and Water Conservation. 23. November-December 1968. pp. 215-218.

37. Wandesforde-Smith, Geoffrey. "The Bureaucratic Response to Environmental Politics." National Resources Journal, II. July 1971. pp. 479-488.

38. Wilson, Thomas W. Jr. International Environmental Action, A Global Survey. University Press, Cambridge, Mass. 1971.

General References

The following references are of general interest for this subject area:

Albertson, Peter, and Barnett, Margery (eds.) Managing the Planet: Essays presented at the International Joint Conference on "Environment and Society in Transition," held at New York, April 27-May 2, 1970. Prentice-Hall, Englewood Cliffs, N.J. 1972. 300 pp.

Antoniou, Jim. Environmental Management. McGraw-Hill, New York. 1972. 192 pp.

Bessey, Roy F. "Environmental Defense and Effective Administration." Public Administration Review. 30. September-October 1970. pp. 563-566.

Bigham, Alastair. Law and Administration Relating to the Environment. J. Whitaker & Sons Ltd., 13 Bedford Sq. London, Oyez Publishing Co., London, England. 1973.

Caldwell, Lynton K. Environmental Studies Series, I, II, III, and IV: Papers presented at a Colloquium on The Politics and Public Administration of Man-Environment Relationships. Sponsored by the Conservation Foundation, held at Indiana University, Bloomington, Indiana, 1965. Institute of Public Administration, Indiana University, Bloomington, Indiana. 1967. various paging:

- Environmental Studies I: Political Dynamics of Control (January 30, 1967). 62 pp.
- Environmental Studies II: Intergovernmental Action on Environmental Policy: The Role of the States (February 28, 1967). 69 pp.
- Environmental Studies III: Politics, Professionalism and the Environment (March 21, 1967). 43 pp.
- Environmental Studies IV: Research on Policy and Administration in Environmental Quality Programs (March 30, 1967). 53 pp.

Caldwell, Lynton K. "Environmental Quality as an Administrative Problem." Annals of the American Academy of Political and Social Science. 391. March 1972. pp. 103-115.

Caldwell and Sidiqui. Environmental Policy, Law, and Administration: A Guide. School of Public and Environmental Affairs, Indiana University, Bloomington, Indiana. June 1974.

Caldwell, L. K. Environment: A Challenge to Modern Society. Doubleday, New York. 1970.

Campbell, Roy R. and Wade, Jerry L. (eds.) Society and Environment: The Coming Collision. Allyn and Bacon, Boston. 1972.

Clement, Thomas M. Jr., and Mountain, Pamela T. Engineering a Victory for Our Environment: A Citizen's Guide to the U.S. Army Corps of Engineers. U.S. Government Printing Office, Washington, D.C. 1972. 413 pp.

Council of Europe. The Management of the Environment in Tomorrow's Europe. European Information Centre for Nature Conservation, Strasbourg, Germany. 1972. 255 pp.

Edmunds, Stahrl W., and Letey, John Jr. Environmental Administration. McGraw - Hill, (McGraw-Hill Series in Management), New York. xv. 1973. 517 pp.

Etzold, David J., and Welsh, Charles A. "Environmental Management": Journal of Environmental Systems, Vol. 1(3). September 1971. pp. 289-302.

E.P.A., Office of Research and Monitoring. An Anthology of Selected Readings for the Symposium on the "Quality of Life" Concept: A Potential New Tool for Decision-Makers. Washington, D.C. 1972.

Ewald, William R. Jr. (ed.). Environment and Policy, The Next Fifty Years. Indiana University Press, Bloomington, Indiana. 1968.

Foss, Phillip O. (ed.). Politics and Ecology. Duxbury Press, Belmont, California. 1972. 298 pp.

Jarrett, Henry (ed.). Comparisons in Resource Management: Six Notable Programs in Other Countries and Their Possible U.S. Application. Published for Resources for the Future by The Johns Hopkins Press, Baltimore. 1961. xv. 271.

Kneese, Allan V. "Strategies for Environmental Management." Public Policy. 19. Winter 1971. pp. 37-52.

MacNeill, J. W. Environmental Management: Constitutional Study Prepared for the Government of Canada. Information Canada. 1971.

National Environmental Information Symposium. Management and Planning. Cincinnati, Ohio. 1972.

New York Academy of Sciences. Public Policy Toward Environment--1973: A Review and Appraisal. New York Academy of Sciences, New York. 1973. 201 pp.

Sheffe, Norman (ed.) Environmental Quality. McGraw-Hill of Canada, Toronto. 1971. 118 pp.

Stein, Robert E. The Potential of Regional Organizations in Managing the Human Environment. Woodrow Wilson International Center for Scholars, Washington, D.C. 1972. 61 pp.

Sumek, Lyle. Environmental Management and Politics: A Selected Bibliography. Center for Government Studies, Northern Illinois University, DeKalb. 1973. 73 pp.

Sweden. Royal Commission on Natural Resources. Environmental Research: Report of the Royal Commission on Natural Resources--Part I: The Area of Research; Part II: Organization and Resources. Sweden Royal Commission on Natural Resources, Stockholm, Sweden. 1968. 93 pp.

United States Congress. House. Committee on Government Operations. Subcommittee on Conservation and Natural Resources. The Environmental Decade (Action Proposals for the 1970's). Hearings of the Subcommittee to the Committee before the 91st Congress, 2d Session, February 2-6, March 13, and April 3, 1970. United States Government Printing Office, Washington, D.C. 1969. 56 pp.

United States Congress. Senate. Committee on Interior and Insular Affairs. A Definition of the Scope of Environmental Management. Report prepared by Daniel Dreyfus for the Committee, and delivered before the 91st Congress, 2d Session, January 1970. United States Government Printing Office, Washington, D.C. 1970. 27 pp.

United States Department of Commerce. Council on Environmental Quality. Environmental Quality: First Annual Report--1970; Second Annual Report--1971; Third Annual Report; and Fourth Annual Report. United States Government Printing Office, Washington, D.C. 1970.

United States Department of Commerce. Council on Environmental Quality. The President's 1971 Environmental Program. United States Government Printing Office, Washington, D.C. 1971. 306 pp.

United States Department of Commerce. Office of Noise Abatement and Control. Summary of Noise Programs in the Federal Government. United States Government Printing Office, Washington, D.C. 1971.

United States President's Office of Science and Technology. Solid Waste Management: A Comprehensive Assessment of Solid Waste Problems, Practices, and Needs. United States Government Printing Office, Washington, D.C. 1969. 111 pp.

Wandesforde-Smith, Geoffrey. "Environmental Administration: Sources of Aid and Comfort?" Public Administration Review. 32. November-December 1972. pp. 881-888.

Woods, Barbara (ed.). Eco-Solutions: A Casebook for the Environment Crisis. Schenkman, Cambridge, Massachusetts. 1972. 540 pp.

ACTOR/ROLE INTERACTIONS

Section IV of part I is a discussion of the nature of actor/role interactions in environmental management. This bibliography was compiled for further development of the issues raised in that section.

Brooks, Harvey. "The Scientific Advisor," from Robert Gilpin and Christopher Wright (ed.). Scientists and National Policy-Making. Columbia University Press, New York and London. 1964. pp. 73-96.

Carter, Luther J. "Earl L. Butz, Counselor for Natural Resources: President's Choice a Surprise for Environmentalists." Science. vol. 179. January 26, 1973. pp. 358-359.

Goldberg, Lesley. "The West Side Highway Story." New Engineer. October 1970. pp. 33-35.

Gillette, Robert. "National Environmental Policy Act: How Well is it Working?" Science. vol. 176. April 14, 1972. pp. 146-150.

Haskell, Elizabeth. "State Governments Tackle Pollution." Environmental Science and Technology. vol. 5. no. 11. November 1971.

Jacobsen, Sally. "Anti-Pollution Backlash in Illinois: Can a Tough Protection Program Survive." Science and Public Affairs. January 1974.

Johnson, Huey J. "Implementation for Goals." In No Deposit--No Return, Man and His Environment: A View Toward Survival. Ed. by Huey D. Johnson, anthology of papers presented at the 13th National Conference of the U.S. National Commission for UNESCO, November 1969, San Francisco, California. Addison-Wesley. 1970.

Krieth, Frank. "Lack of Impact." Environment. vol. 15. no. 1. January/February 1973. pp. 26-33.

Laserson, Nina. "Nine Reasons to Take the Consumerist Seriously." Innovation. November 1971. Number twenty-six.

Like, Irving. "Multi-Media Confrontation--The Environmentalists' Strategy for a No-Win Agency Proceeding." Ecology Law Quarterly. vol. 1. no. 495. 1971. pp. 495-518.

MacNeill, J. W. Environmental Management. Constitutional study prepared for the Government of Canada. Information Canada, Ottowa. 1971.

McManimie, Robert, and McGregory, William. "Business Synergism." Innovation. November 1971. Number twenty-six.

Perl, Martin L. "The Scientific Advisory System: Some Observations." Science. vol. 177. December 29, 1972. pp. 1166-1171.

Primack, Joel and Von Hipple, Frank. "Public Interest Science." Science. vol. 177. December 29, 1972. pp. 1166-1171.

Richards, Thomas W. "Clearing Obstacles and Getting Results at the Local Level." In No Deposit--No Return, Man and His Environment: A View Toward Survival. Ed. by Huey D. Johnson, anthology of papers presented at the 13th National Conference of the U.S. National Commission for UNESCO, November, 1969, San Francisco, California. Addison-Wesley. 1970.

U.S. Council on Environmental Quality. Environmental Quality. U.S. Government Printing Office, Washington, D.C. 1973.

U.S. Environmental Protection Agency, Office of Research and Monitoring, Environmental Studies Division. The Quality of Life Concept: A Potential New Tool for Decision-Makers. Report of a symposium held August 29-31, 1972.

Wilson, Thomas W. Jr. International Environmental Action: A Global Survey. Dunellen Univ. Press of Cambridge, Mass. 1971.

Graik, Kenneth H. "The Environmental Dispositions of Environmental Decision-Makers." The Annals of the American Academy of Political and Social Science. 389. May 1970. pp. 87-94.

Kaynor, E. R. and Howards, Irving. "Attitudes, Values, and Perceptions in Water Resource Decision-Making within a Metropolitan Area." Publication No. 29, Water Resources Research Center, University of Massachusetts at Amherst. June 1973.

/301.3107M442R>C1/